Genetics and the Extinction of Species

Genetics

and

the Extinction of Species

DNA AND THE CONSERVATION OF BIODIVERSITY

Laura F. Landweber and Andrew P. Dobson, Editors

PRINCETON UNIVERSITY PRESS

PRINCETON, NEW JERSEY

Copyright ©1999 by Princeton University Press
Published by Princeton University Press, 41 William Street,
Princeton, NJ 08540
In the United Kingdom: Princeton University Press,
Chichester, West Sussex

Library of Congress Cataloging-in-Publication Data

Genetics and the extinction of species : DNA and the conservation of biodiversity /
Laura F. Landweber and Andrew P. Dobson, editors.
p. cm.
Includes bibliographic references and index.
ISBN 0-691-00970-8 (cl: alk. paper). – ISBN 0-691-00971-6 (pb : alk. paper)
1. Conservation biology. 2. Population genetics. I. Landweber, Laura F.
(Laura Faye), 1967- . II. Dobson, Andrew P.
QH75.G454 1999
  576.5'8–dc21   99-24114
                                                  CIP

This book has been composed in Galliard using LaTeX.

The paper used in this publication meets the minimum
requirements of ANSI/NISO Z39.48-1992(R 1997) (Permanence of Paper)
http://pup.princeton.edu
Printed in the United States of America
1 3 5 7 9 10 8 6 4 2
1 3 5 7 9 10 8 6 4 2

Designed by Thomas R. Hagedorn

Contents

CONTRIBUTORS

William Amos, Department of Zoology, Downing Street, Cambridge CB2 3EJ, UK

Rebecca L. Cann, University of Hawaii at Manoa, John A. Burns School of Medicine, Department of Genetics and Molecular Biology, 1960 East-West Road, Honolulu, HI 96822, USA

Kathryn M. Rodríguez-Clark, Department of Ecology and Evolutionary Biology, Princeton University, Princeton, NJ 08544-1003, USA

Andrew P. Dobson, Department of Ecology and Evolutionary Biology, Princeton University, Princeton, NJ 08544-1003, USA

Leslie J. Douglas, University of Hawaii at Manoa, John A. Burns School of Medicine, Department of Genetics and Molecular Biology, 1960 East-West Road, Honolulu, HI 96822, USA

Leonard A. Freed, Department of Zoology, University of Hawaii, Honolulu, HI 96822, USA

Paul H. Harvey, Department of Zoology, University of Oxford, South Parks Road, Oxford OX1 3PS, UK

Kent E. Holsinger, Department of Ecology and Evolutionary Biology, University of Connecticut, Storrs, CT 06269-3043, USA

Russell Lande, Department of Biology, University of Oregon, Eugene, OR 97403-1210, USA

Laura F. Landweber, Department of Ecology and Evolutionary Biology, Princeton University, Princeton, NJ, 08544-1003, USA

Roberta J. Mason-Gamer, Department of Biological Sciences, University of Idaho, Moscow, ID 83844-3051, USA

Helen Steers, Department of Zoology, University of Oxford, South Parks Road, Oxford OX1 3PS, UK

Jeannette Whitton, Department of Botany, University of British Columbia, Vancouver, BC V6T 1Z4, Canada

ILLUSTRATIONS

Figures

Tables

PREFACE

The chapters in this volume are based upon talks given at the symposium entitled, "Genes, species, and the threat of extinction: DNA and genetics in the conservation of endangered species," held at Princeton University on October 4, 1996 in honor of Princeton's Bicenquinquagenary. We are very grateful to the university, particularly to the 250th Anniversary fund, and to the Jane H. Fortune Distinguished Lectureship in Conservation Biology for providing funds for the symposium. The editors would also like to express their thanks to Tom Hagedorn for his help in producing this volume.

Princeton, New Jersey
February, 1999

Introduction: Genetics and Conservation Biology

ANDREW P. DOBSON

Population genetics has consistently played a central role in the development of conservation biology as a science. All species are divided into populations; Hughes, Daily, and Ehrlich (1997) recently attempted to estimate the number of "populations" currently inhabiting Earth. They estimate that if each species is divided into approximately 220 populations, then Earth is occupied by between 1.1 and 6.6 billion populations. Unfortunately, if we assume that population extinction is a simple linear function of habitat loss, then we are losing 1,800 populations per hour (16 million annually). As each of these populations has the potential to evolve into a new species, then we are plainly massively inhibiting the ability of Earth's inhabitants to produce the diversity that will allow it to cope with future changes in climate, or other long-term threats.

The realization that humans are dependent on the resources provided by other species has provided a tremendous incentive for the development of conservation biology over the last ten years. Yet as conservation biology becomes more oriented toward economic, philosophical, and policy considerations, it could seem as though genetics plays a less crucial role.

This volume presents a set of papers whose contents refute this naive conjecture. The topics covered in the book range from a comprehensive reexamination of the interaction between genetics, demography, and different types of stochasticity, through a detailed overview of the role that genetics plays in captive breeding schemes, to a number of empirical studies that sharply outline insights into population structure that can only be provided by detailed examination of the factors that produce and maintain genetic variability in natural populations. The chapters illustrate that population genetics remains a cornerstone of conservation biology. Furthermore, the technological and theoretical developments that have taken place over the last ten years have provided a comprehensive new understanding of the genetic structure of natural populations. This has allowed important new insights to emerge about how populations respond to past selection pressures. If conservation biology is to continue to progress as a scientific discipline that influences environmental policy, it has to combine up-to-date, original, and solidly argued science with insightful advocacy.

Part of the motivation for organizing the symposium that produced the current volume was a chance remark at the annual general meeting of the Society for Conservation Biology. The editor of the society's journal noted that the rejection rate for articles on population genetics was much higher than that for other topics. His only explanation was that the geneticists probably had higher standards than referees from other sub-disciplines of conservation biology. This remark simultaneously delights and concerns me. It is obviously excellent news that standards of refereeing are high; this effectively ensures that only first-rate papers will be published in the area of genetics. It worries me deeply that it is easier for authors in other sections of the discipline to publish papers that are more speculative and less firmly based on solid science and sound reasoning. This will only encourage young people to enter the policy arena with weaker scientific training than they will need to support (or even comprehend) the scientific implications of the policies they develop.

The role that inbreeding played in species endangerment was one of the major initial areas of focus for conservation biologists. Although we still appreciate the potential importance of inbreeding in captive populations, we increasingly recognize that by the time inbreeding depression is important in the wild, it is likely that populations will decline to extinction for simple demographic reasons. While there may be some synergistic interaction between inbreeding depression and the stochastic extinction of small populations, it is unlikely that inbreeding depression per se was important in the early stages of the decline. However, conservation biologists who study endangered and threatened populations have developed a whole range of other genetic techniques in the last ten to twenty years. The main purpose of this book is to provide an introduction to some of these new and exciting areas and to illustrate the novel roles that genetics can play in understanding and managing biodiversity.

In this volume we have taken the central theme of Graham Caughley's (1994) masterly critique of conservation biology to heart. Caughley suggests that conservation biologists have paid too much attention to small populations that are close to extinction, while ignoring the major factors that led to their initial decline. We have therefore tried to establish a balance between papers that provide new insights into the genetics of small populations and those that examine how new genetic techniques can be applied to examine populations in the earlier stages of decline.

We envision the book could be used as the basis of a graduate or upper-level class on the role of population genetics in conservation biology. It might also be used as a brief primer to bring the rushed policymaker (with a biological background) up to date on the latest developments in the area. Like all symposia volumes, it is not intended to be comprehensive; instead it highlights areas we view as particularly salient or promising for future research. We have thus tried to establish a balance between chapters that provide a comprehensive introduction to the key areas where population genetics has influenced conservation biology and chapters that describe new approaches to conservation and evolutionary genetics. Finally, we have included several chapters that we hope will provoke the reader; such a response may lead to the development of new insights into the importance of genetics in long-term conservation planning.

A number of excellent volumes have recently appeared that complement the work presented in this volume. In particular, the volume by Avise and Hamrick (1996) provides an excellent collection of field studies from a broad taxonomic spectrum. Similarly, the volume edited by Loeschcke, Tomack, and Jain (1994) provides a comprehensive conceptual introduction to the whole subject, while also illustrating a range of insightful examples. It is still well worth examining the two volumes that essentially initiated this whole research enterprise: Frankel and Soulé (1981) and Schonewald-Cox, Chambers, MacBryde, and Thomas (1983).

To be comprehensive, we acknowledge there are several areas that were not part of this symposium. For essays on the use of modern genetic methods to reconstruct the historical geographic distribution of a species or community whose natural habitat was converted to agricultural or urban areas, we refer the reader to Avise and Hamrick (1996). Similarly, there is no chapter that deals specifically with the interaction between the erosion of genetic diversity and changes in disease susceptibility, but this subject was reviewed by Hedrick (1992). Finally, we acknowledge that there is no chapter that deals with the many recent contributions that genetic techniques have made to systematics and taxonomy; indeed, modern taxonomic techniques have revolutionized systematics! Here we would strongly recommend the volume *Systematics and Conservation Evaluation,* edited by Forey, Humphries, and Vane-Wright (1994). However, the present book's Chapter 5, by Harvey and Steers, describes some new methods for the analysis and interpretation of the gene sequence data used to construct phylogenies. This work provides an introduction to a fascinating new set of techniques for examining the past evolutionary trajectories of species for which appropriate genetic sequence data are available.

The main body of the book commences with a chapter by Russell Lande that introduces the major threats to endangered and threatened species. All of these reduce the size and viability of natural populations and produce a cascade of effects that threaten their ecological and evolutionary potential. The chapter then goes on to lay to rest a result that has plagued conservation biology for the last twenty years. In the early years of the discipline, Lande was attributed with the suggestion that a simple rule of thumb provided guidelines for the minimum size of a population that would ensure its long-term demographic (50 individuals) and evolutionary viability (500 individuals) (Lande 1976; Franklin 1980). Unfortunately, the figures were an approximation based on estimates of mutation rate for neutral alleles in one species of *Drosophila*. This estimation ignored a whole range of important biological details from selection through to interspecific variability in mutation rate. Nevertheless, the beguiling simplicity of the 50/500 "rule" was so appealing that it very nearly became a part of the legal classification of a species as endangered or on the route to recovery. After illustrating the problems with this approximation, Lande's chapter goes on to explore the interaction between demography and population dynamics. It then provides a clear and insightful overview of the economic conflicts that arise when we attempt to exploit species. The chapter sharply illustrates that the development of rules that guide effective conservation policy requires an insightful interdisciplinary synthesis of really quite complex ideas, rather than application of simple precepts.

Chapter 2, by Holsinger, Mason-Gamer, and Whitton, provides an insightful set of examples of plant conservation. Developing some of the themes described in more detail in the book edited by Falk and Holsinger (1991), *Genetics and the Conservation of Rare Plants*, this chapter illustrates two important general principles. First, loss of variability is indeed a symptom rather than a cause of endangerment. Second, the chapter illustrates that caution is essential in interpreting data based on different types of genetic markers, as such markers may be associated with genotypes that are completely different from those involved in some future response to selection.

Kathryn Rodríguez-Clark's chapter (Chapter 3) provides an insightful introduction to the genetic methods used in mate choice in captive breeding programs. The chapter argues that breeding plans should be based upon mean kinship, rather than such measures as genetic uniqueness, founder importance coefficients, target founder contribution, or founder genome equivalents. This is followed by Bill Amos's chapter (Chapter 4), which provides a critical discussion of two important problems in conservation genetics. First, can observed levels of genetic variation in nature be used as reliable indicators of "genetic health"? Second, can microsatellite variation be used as a measure of interpopulation genetic differentiation?

As mentioned earlier, Harvey and Steers's chapter (Chapter 5) describes some new methods for the analysis and interpretation of the gene sequence data which are usually used to construct phylogenies. This work provides an introduction to a fascinating new set of techniques that may be used to examine the past evolutionary trajectories of species. It also allows us to begin to ask the most worrying question in conservation biology: "How much diversity can we lose before evolution stops?" Sean Nee and Robert May have addressed this question in a paper that appeared shortly after the symposium (Nee and May 1997). The reassuring answer is that approximately 80 percent of the underlying trees of life can survive even when 95 percent of species are lost. These papers provide an important new perspective on the current debate about the role that systematics might play in setting priorities for conservation.

We then have two chapters (Chapters 6 and 7) that examine in detail the plight of one particular group of endangered species, the native Hawaiian avifauna. These birds present a classic example of an isolated radiation that has given rise to a closely related but morphologically very distinct group of species. They are threatened by a range of factors, from the massive loss of natural habitat from lowland areas in Hawaii, to the introduction of feral pigs and goats. These threats were further exasperated by the introduction of alien bird species, with the naive hope of providing an aesthetic substitute for the declining native avifauna! Sadly, the introduced species were hosts to avian malaria and this led to further decline in the native bird species. The two papers by Rebecca Cann and Leslie Douglas (Chapter 6) and by Leonard Freed (Chapter 7) provide a comprehensive overview of the anthropogenic and biological threats experienced by the Hawaiian birds and the specific threats posed by avian malaria. The chapter by Cann and Douglas illustrates how modern molecular techniques provide important epidemiological information that allows one to trace the routes of transmission from important reservoir hosts. The chapter by Freed illustrates how comparative approaches provide insights into subtle differences in the threats to each species.

The volume concludes with a chapter by Laura Landweber (Chapter 8) which describes how modern genetic techniques may be applied to extract and examine the DNA of individuals that have been dead for many years, and even of species that have been extinct for many decades or centuries. These techniques provide an important way of examining some of the components of biodiversity that may have been lost as a direct cost of the expansion of the human exercise. A fervant technocrat might argue that these techniques provide us with a way of eventually reconstructing and potentially rescuing any vital components of biodiversity that might get lost in the course of the current massive anthropological expansion. However, this chapter and all the others in the book emphasize that the most effective ecological and economic way to preserve biodiversity is to provide and maintain the natural areas where species evolved in the first place. While the book celebrates the increased understanding of natural variability that new genetic techniques have provided in the last ten years, all of the authors retain a deep worry that there will be considerably less biodiversity when Princeton University celebrates its 300th Anniversary in less than fifty years time.

REFERENCES

Avise, J. C., and Hamrick, J. L., eds. (1996). *Conservation Genetics, Case Histories from Nature.* New York: Chapman and Hall.

Caughley, G. (1994). Directions in conservation biology. *J. of Animal Ecol., 63,* 215–244.

Falk, D. A., and Holsinger, K. E., eds. (1991). *Genetics and the conservation of rare plants.* Oxford: Oxford University Press.

Forey, P. L., Humphries, C. J., and Vane-Wright, R. I., eds. (1994). *Sytematics and Conservation Evaluation.* The Systematics Association Special Volumes. Oxford: Clarendon Press.

Frankel, O. H., and Soulé, M. E. (1981). *Conservation and Evolution.* Cambridge: Cambridge University Press.

Franklin, I. R. (1980). Evolutionary change in small populations. In Soulé, M. E., and Wilcox, B. A. (eds.), *Conservation Biology. An Evolutionary-Ecological Perspective,* 135–150. Sunderland, Mass.: Sinauer.

Hedrick, P. W. (1992). Conservation Genetics: Techniques and Fundamentals, *Ecol. Appl., 2,* 30–46.

Hughes, J. B., Daily, G. C., and Ehrlich, P. R. (1997). Population diversity: Its Extent and Extinction. *Science, 278,* 689–691.

Lande, R. (1976). The maintanence of genetic variability by mutation in a polygenic character with linked loci. *Genet. Res. Camb., 26,* 221–35.

Loeschcke, V., Tomiuk, J., and Jain, S. K., eds. (1994). *Conservation Genetics.* Boston: Birkhauser Verlag.

Nee, S., and May, R. M. (1997). Extinction and the loss of evolutionary history. *Science, 278,* 692–694.

Schonewald-Cox, C. M., Chambers, S. M., MacBryde, B., and Thomas, W. L., eds. (1983). *Genetics and Conservation: A reference for managing wild animal and plant populations.* London: Benjamin-Cummings.

1

Extinction Risks from Anthropogenic, Ecological, and Genetic Factors

RUSSELL LANDE

SUMMARY. This chapter discusses the effects of both deterministic and stochastic factors on the risk of extinction. I begin by introducing the anthropogenic factors, such as land development, overexploitation, species translocations and introductions, and pollution, that are the primary causes of endangerment and extinction. These primary anthropogenic factors have ramifying ecological and genetic effects that contribute to extinction risk. I then discuss the role of ecological factors, including environmental stochasticity, random catastrophes, and metapopulation dynamics (local extinction and colonization). Genetic factors, such as hybridization with nonadapted gene pools and selective breeding and harvesting, also play a critical role. Especially important in small populations are the genetic factors of inbreeding depression, loss of genetic variability, and fixation of new deleterious mutations, as well as the ecological factors of Allee effect, edge effects, and demographic stochasticity. Finally, I consider the relative importance and interaction of these different risk factors, as they affect population dynamics and the threat of extinction.

INTRODUCTION

Many plant and animal species around the world are threatened or endangered with extinction, largely as a result of human activities. The frequent multiplicity of threatened and endangered species, even within local planning areas, has made it clear that effective conservation and restoration must be done in the context of comprehensive landscape and

An earlier version of this chapter was published as: Lande, R. (1998) Anthropogenic, ecological and genetic factors in extinction. In *Conservation in a Changing World*: 29-51. Mace, G. M., Balmford, A., and Ginsberg, J. (eds.). Cambridge: Cambridge University Press.

ecosystem approaches that consider biodiversity and large-scale ecological processes. Species-based approaches should nevertheless play an essential role in formulating and monitoring large-scale conservation and restoration plans to ensure that ecologically important species, or those that indicate ecosystem health, are properly managed. Understanding the factors that contribute to the extinction risk of particular species, therefore, remains of critical importance even within landscape and ecosystem approaches to conservation and restoration.

Anthropogenic factors constitute the primary causes of endangerment and extinction: land development, overexploitation, species translocations and introductions, and pollution. These primary factors have ramifying ecological, and genetic effects that contribute to extinction risk. For example, land development causes habitat fragmentation, isolation of small populations, and intensification of metapopulation dynamics. All factors affecting extinction risk are ultimately expressed, and can be evaluated, through their operation on population dynamics. Here I review anthropogenic, ecological and genetic factors contributing to extinction risk, briefly discussing their relative importance and interactions in the context of conservation planning.

1.1 ANTHROPOGENIC FACTORS

Land Development

Human population growth and economic activity convert vast areas for settlement, agriculture, and forestry. This results in the ecological effects of habitat destruction, degradation, and fragmentation, which are among the most important causes of species declines and extinctions. Habitat destruction contributes to extinction risk of three-quarters of the threatened mammals of Australasia and the Americas and more than half of the endangered birds of the world (Groombridge 1992, ch. 17).

Overexploitation

Unregulated economic competition

Inadequately regulated competition among resources extractors, especially in open-access fisheries and forestry, is one of the major causes of resource overexploitation and depletion (Ludwig, Hilborn, and Walters 1993; Rosenberg *et al.* 1993). About half of the fisheries in Europe and the United States were recently classified as overexploited (Rosenberg *et al.* 1993). Hunting and international trade contributes to the extinction risk of over half of the threatened mammals of Australasia and the Americas and over one-third of the threatened birds of the world (Groombridge 1992, ch. 17) and has caused local extinctions of many forest-dwelling mammals and birds even in areas where habitat is largely intact (Redford 1992).

Economic discounting

A nearly universal economic practice is the discounting of future profits. Annual discount rates employed by many governments and resource exploiters are often in the range of 5% to 10% or higher. Clark (1973, 1990) showed that in many cases there is a critical discount rate above which the optimal strategy from a narrow economic viewpoint is immediate harvesting of the population to extinction (liquidation of the resource). In simple deterministic models with a constant profit per individual harvested, the critical discount rate equals the maximum per capita rate of population growth, r_{max}, because money in the bank grows faster than the population (May 1976). Organisms with long generation time and/or low fecundity, such as many species of trees, parrots, sea turtles, and whales, have r_{max} below the prevailing discount rate and are frequently threatened by overexploitation.

Stochastic fluctuations in population size reduce sustainable harvests (Beddington and May 1977; Lande, Engen, and Sæther 1995). Optimal harvesting strategies that reduce extinction risk as well as maximize sustainable harvests have only recently been developed. Such harvesting strategies generally involve threshold population sizes below which no harvesting occurs when the population fluctuates below the threshold (allowing the population to increase at the maximum natural rate), and above which harvesting occurs as fast as possible (Lande, Engen, and Sæther 1994, 1995; Lande, Sæther, and Engen 1997).

Introduction of Exotic Species

Numerous species are transported and released in foreign environments both accidentally and deliberately in private and commercial transportation, live-animal trade, ornamental plantings, and biological control. Introduced species, mainly predators and competitors, seriously affect about one-fifth of the endangered mammals of Australasia and the Americas and birds of the world (Groombridge 1992, ch. 17). Introduced rats are responsible for extinctions of many island-endemic birds (Atkinson 1989). In some national parks in Hawaii, up to half of the plant species are nonnative (Vitousek 1988) and constitute a serious risk for the endangered flora. Introduced strains and species of parasites and diseases also pose a serious problem for many endangered species (Dobson and May 1986).

Pollution

Agricultural and industrial pollution have had both localized and widespread effects. Long-lasting pesticides, such as DDT, become concentrated in terrestrial and aquatic food chains, and have endangered several birds of prey, such as the American bald eagle and peregrine falcon. Although bans on most long-lasting pesticides in the United States helped recovery of both these species, the pesticides are still used in many countries. About 4% of endangered birds of the world and 2.5% of mammals of Australasia and the Americas are at risk from pollution (Groombridge 1992, ch. 17). These figures underestimate the extent of morbidity, mortality, and fertility impairment caused by pesticides in many non-endangered species.

Acid rain has had intense regional effects on terrestrial plant communities in Western Europe and on freshwater ecosystems in the eastern United States. In Germany, about one-fourth of the native species of ferns and flowering plants are endangered or extinct, with about 5% affected by air and soil pollution and 5% by water pollution (Organization for Economic Cooperation and Development [ODEC] 1991).

1.2 Ecological Factors

Environmental Fluctuations and Catastrophes

Unexploited vertebrate populations fluctuate through time with coefficients of variation (standard deviation/mean) in annual abundance usually in the range of 20% to 80% or more (Pimm 1991). Exploited populations also are highly variable (Myers, Bridson, and Barrowman 1995), due not only to environmental stochasticity, but also because commonly used exploitation strategies, such as constant-effort or constant-rate harvesting, tend to reduce population stability (Beddington and May 1977; May *et al.* 1978). Catastrophes are an extreme form of environmental fluctuation in which the population is suddenly reduced in size by a large proportion, usually because of extraordinary climatic conditions such as droughts or severe cold or because of disease outbreaks (Young 1994).

In stochastic population models with density-dependence and either normal environmental fluctuations in population growth rate or random

Risk Factor	_Proportional scaling of T_
Declining population[1]	$-(\ln N)/\bar{r}$
Environmental stochasticity[2]	$N^{2\bar{r}/V_e - 1}$
Demographic stochasticity[3]	$(1/N)e^{2Nr/V_1}$
Fixation of new mutations[4]	N_e^{1+1/c^2}

[1] In this case only, mean population growth rate, $\bar{r}$, is negative.

[2] Mean and variance of annual growth rate are $\bar{r}$, V_e respectively.

[3] Mean and variance of individual Malthusian fitness are r, V_1, respectively.

[4] c is the coefficient of variation of selection against new mutations.

Table 1.1. Asymptotic scaling laws for mean time to extinction, T, from different ecological and genetic risk factors, as a function of the initial actual size, N, or effective size, N_e, of a population at carrying capacity.

catastrophes, for a population initially near carrying capacity the mean time to extinction, T, scales asymptotically (for sufficiently large populations) as the carrying capacity raised to a power. Depending on the magnitude of environmental fluctuations in population growth rate, or on the frequency and magnitude of catastrophes, this power may be either greater than or less than one. Thus, comparing populations with different carrying capacities, under low environmental stochasticity, T increases faster than linearly with increasing carrying capacity, whereas under high environmental stochasticity, T increases less than linearly with increasing carrying capacity (Lande 1993). Logarithmic scaling of T with initial population size is characteristic of a declining population, in which the mean growth rate is negative, regardless of whether the decline is deterministic or stochastic (Lande 1993). Asymptotic scaling laws for various risk factors are summarized in Table 1.1.

If population subdivision substantially reduces the correlation in environmental stochasticity among localities, for example, considering one large contiguous reserve versus several small distant reserves of the same total area, then subdivision can increase T. Thus, in a case when single populations are subject to major catastrophes, occurring randomly among populations, then population subdivision can clearly be advantageous for persistence (Burkey 1989). Subdivision also can increase persistence in the presence of catastrophic epidemics, not only by reducing the transmission of epidemics among localities, but in some cases by reducing their frequency because many epidemics require a threshold population size or density to become started (Hess 1996). On the other hand, if the small reserves are located in the same general area, and are subject to nearly the same environmental stochasticity, then subdivision is likely to decrease persistence by making the small reserves more subject to edge effects, Allee effects, and demographic stochasticity.

Above the species level, ecosystem function is likely to be enhanced in large reserves without landscape fragmentation; species diversity would tend to be larger, at least initially, in several small reserves spread over a larger geographic area, but these would suffer more rapid local extinctions. Designs for nature reserves systems must balance these advantages and disadvantages of subdivision. From a review of data on several natural and artificial archipelagoes, Burkey (1995) concluded that on a single large island the rate of species extinctions is initially faster, but ultimately slower, than on several small islands.

Metapopulation Dynamics

Dispersal among local populations, patches of suitable habitat or "islands" also can have advantages and disadvantages for persistence. The major advantage is that dispersers can recolonize suitable habitat after local extinctions, allowing a metapopulation to persist in a balance between local extinction and recolonization (Levins 1970; Hanski and Gilpin 1997). In Levins's (1970) original metapopulation model, the equilibrium proportion of islands occupied by a species is $p = 1 - e/m$, where e is the rate of local extinction and m is the colonization rate. Metapopulation persistence then requires that $m > e$. This and other demographic and genetic benefits of dispersal (described below) have spurred interest in various methods of enhancing dispersal, from artificial transport of individuals or germ cells to preservation or creation of habitat corridors connecting islands of suitable habitat. The functioning of corridors has hardly been tested (Andreassen, Hall, and Ims 1996; Downes, Handasyde, and Elgar 1997), but for many species they may be of little value because of edge effects.

The major disadvantage to dispersal occurs during individual movement through unsuitable habitat in a heterogeneous landscape. For species in which individuals or mated pairs hold exclusive territories, or home ranges with small overlap, the basic effects of habitat destruction and fragmentation can be taken into account, along with life history, age-structured population dynamics, and individual dispersal behavior (Lande 1987). Identifying the individual territory as the local unit in a metapopulation, local extinction corresponds to the death of an individual, and colonization represents successful dispersal into a suitable, unoccupied patch of habitat. Patches of suitable habitat the size of individual territories are assumed to be randomly or evenly distributed across a large region, with no clumping on a spatial scale much larger than the mean individual dispersal distance, and the proportion of the region composed of suitable habitat is h. The equilibrium proportion of suitable habitat occupied by the species is $p = 1 - (1 - k)/h$. The "demographic potential," k, depends on the life history and dispersal behavior of the species and generally lies between 0 and 1; it gives the maximum occupancy of suitable habitat in a completely suitable region ($p = k$ when $h = 1$).

This model reveals two general and robust features important for conservation planning. First, unoccupied suitable habitat may be as important as occupied habitat for long-term persistence of a metapopulation. Continual destruction of unoccupied habitat will doom a metapopulation to early extinction. Second, as the amount of suitable habitat in a region decreases by habitat destruction and fragmentation, the equilibrium occupancy of suitable habitat decreases. Since the population size equals the amount of suitable habitat multiplied by the occupancy of suitable habitat, this implies that with habitat destruction and fragmentation the equilibrium population size generally declines faster than the rate of habitat loss. This implies the existence of an "extinction threshold" or minimum density of suitable habitat in a region necessary for population persistence, at $h = 1 - k$ (since $p = 0$ when $h \leq 1 - k$). A population may become extinct in the presence of a large amount of suitable habitat if this is too sparsely distributed, as first shown by application of this model to the northern spotted owl (Lande 1988a; Doak 1989; Thomas *et al.* 1990; McKelvey *et al.* 1993).

Few attempts have been made to analyze the effects of dispersal and local population dynamics on metapopulation persistence that could be applied more generally to nonterritorial species. Initial work in this direction (Hanski and Gyllenberg 1993; Hanski *et al.* 1995; Lande, Engen, and Sæther 1998) suggests the existence of alternative equilibria, a stable equilibrium with high occupancy of suitable habitat, and an unstable equilibrium with low occupancy of suitable habitat. These alternative equilibria arise because of interactions between local and global population dynamics. At low-habitat occupancy, emigration from an occupied patch is not compensated by immigration, which can render isolated populations vulnerable to extinction. Increasing habitat occupancy in the metapopulation increases the number of immigrants to any site, decreasing the rate of local extinction (the "rescue effect" of Brown and Kodric-Brown 1977) and increasing the probability of successful colonization ("establishment effect" of Lande, Engen, and Sæther 1998). The unstable equilibrium at low-habitat occupancy constitutes a kind of Allee effect at the metapopulation level.

Small Population Size

Demographic Stochasticity

Random individual variation in vital rates of mortality and reproduction, and random variation in adult sex-ratio, cause fluctuations in the per capita growth rate of small populations. The magnitude of these fluctuations are inversely proportional to population size because independent random events among individuals tend to average out in a large population. In contrast to environmental stochasticity, which may operate with equal intensity in populations of any size, demographic stochasticity affects small populations most strongly. Demographic stochasticity is generally thought to be of relatively little importance in populations larger than roughly 100 individuals (MacArthur and Wilson 1967; Richter-Dyn and Goel 1972; Lande 1993). In small populations demographic stochasticity may be the dominant stochastic factor in population dynamics, posing a greater risk of extinction than environmental stochasticity. For a population initially at carrying capacity, under demographic stochasticity alone the mean time to extinction scales asymptotically almost exponentially with carrying capacity (Table 1.1). Demographic stochasticity can create a type of Allee effect such that in populations below a certain small size, most population trajectories tend to decrease, resulting in a high probability of extinction (Lande 1998).

Allee effect

In populations below a certain size or density, individuals may suffer reduced fitness from insufficient cooperative interactions with conspecifics. Cooperative social interactions occur in many animal species, including group defense against predators, physical or chemical conditioning of the environment (e.g., huddling for warmth), communal nesting, and increased per capita efficiency of group foraging. More generally, in small or sparsely distributed populations, individuals may have difficulty encountering potential mates. These effects can render population growth negative in small populations, creating an unstable equilibrium at small population size below which the population tends to decline to extinction (Allee *et al.* 1949; Andrewartha and Birch 1954). For example, the Lakeside Daisy is a self-incompatible perennial, and the last individuals in Illinois were found to be incompatible and hence incapable of reproduction (DeMauro 1993).

Edge effects

Habitat destruction and fragmentation create new edges between habitat types, and may reduce habitat quality for considerable distances inside suitable habitat patches, by causing microclimatic alterations and facilitating incursion or invasion of exotic species. For example, clearing tropical rainforests for pastureland causes desiccation and vegetational changes up to hundreds of meters inside remnant forest patches (Lovejoy *et al.* 1986). Fragmentation of temperate-zone forests by agriculture and settlement facilitates the invasion of cowbirds that parasitize the nests of other birds, some of which are endangered (Robinson *et al.* 1995).

Another type of edge effect arises from dispersal beyond the boundary of suitable habitat. The rate of dispersal into unsuitable regions determines the minimum size of a geographically isolated patch of suitable habitat that can support a stable population, known as the critical patch size. With random dispersal, lethal surroundings, and a low intrinsic rate of increase per generation, the critical patch size is much larger than the average individual dispersal distance (Kierstead and Slobodkin 1953). More hospitable surroundings, high intrinsic rate of increase, and habitat selection behavior decrease the critical patch size (Okubo 1980; Pease, Lande, and Bull 1989).

1.3 GENETIC FACTORS

Maladaptive Translocation and Hybridization

A low rate of interspecific hybridization often occurs between closely related species and may be beneficial in augmenting intraspecific genetic variance and adaptive evolution (Lewontin and Birch 1966; Grant and Price 1981). Artificial disturbance of habitats facilitates contact and hybridization between normally noninterbreeding species. Interspecific contact and hybridization also can occur through invasion or introduction of exotic species. Abnormally high rates of interspecific hybridization are likely to be maladaptive because of partial sterility and reduced viability caused by a variety of postzygotic reproductive isolating mechanisms that usually exist between species (Dobzhansky 1970), which can exert a heavy cost on a rare species hybridizing with a common species (Levin, Francisco-Ortega, and Jansen 1996). When natural reproductive isolating mechanisms are mainly prezygotic, interspecific hybridization may not threaten the demographic stability of a species but can nevertheless destroy its genetic integrity. To give some examples, fragmentation of old-growth forests in the Pacific Northwest of the United States has facilitated range extension of the barred owl, which is now hybridizing with the northern spotted owl. Molecular genetic evidence indicates that domestic dogs are hybridizing with the endangered Simien jackal (Wayne 1996). The diversity of subspecies and species of cichlid fish in Lake Victoria has declined with artificial increases in water turbidity that decreases color vision, sexual selection, and mate choice (Seehausen, van Alphen, and Witte 1997).

Intraspecific hybridization also can produce maladaptive effects by eroding the genetic basis of adaptations to local environmental conditions. This often occurs when nonlocal genetic strains are used for restocking game fish and forest trees. Until recently, little attention was given to the genetic properties of introduced stocks, which has resulted in widespread decreases in fitness of stocked populations and maladaptive hybridization with remaining wild stocks. This is one of the major factors contributing to massive declines and numerous local extinctions of salmon runs in the Pacific Northwest (Nehlsen, Williams, and Lichatowich 1991; Ratner, Lande, and Roper 1997).

Selective Breeding and Harvesting

Exploited populations frequently are subject to intense selective harvesting based on individual size, age, and behavior. This can induce evolutionary changes in life history and social structure, which will usually decrease the fitness of the wild population as well as diminish the quantity and quality of future harvests. Selective harvesting is thought to be a factor in body-size declines of many exploited stocks of anadromous fish (Stokes, McGlade, and Law 1993).

Intense selective pressures also can occur during artificial propagation of captive populations, because artificial environments usually differ substantially from natural ones (Arnold 1995). The resulting evolutionary changes are again likely to be maladaptive for populations reintroduced into the wild. Artificial propagation for the purposes of augmentation and reintroduction should be done in as few generations under as naturalistic conditions as possible. Indefinite restocking programs, such as fish hatcheries, may be doing more harm than good in the long run, and should not be viewed as an adequate substitute for habitat restoration (Allendorf and Waples 1996).

Small Population Size

Inbreeding depression

Matings between closely related individuals tend to produce offspring with reduced fitness due to the expression of (partially) recessive deleterious mutations in homozygous form. In historically large, outcrossing populations this inbreeding depression in fitness typically produces a loss of fitness of a few to several percentages per 10% increase in the coefficient of inbreeding or consanguinity (Ralls and Ballou 1983; Falconer and Mackay 1996). Thus, in species of domesticated animals, experimental propagation by continued brother-sister matings generally results in extinction of a high proportion of lines within five or ten generations (Soulé 1980; Frankham 1995a). Species, and populations within a species, differ substantially in the magnitude of inbreeding depression (Soulé 1980; Lacy, Petric, and Warneke 1993).

The genetic basis of inbreeding depression is best understood in species of *Drosophila*, in which roughly equal parts are contributed by nearly recessive lethal mutations and by partially recessive (nearly additive) mildly deleterious mutations (Simmons and Crow 1977). Both recessive lethal and mildly deleterious mutations arise at thousands of genetic loci throughout the genome in eukaryotic species (Simmons and Crow 1977). Theory and experiment indicate that gradual inbreeding allows natural selection to purge recessive lethal mutations from a population as they become expressed in homozygotes, whereas it is difficult or impossible for inbreeding to purge the more nearly additive mildly deleterious mutations (Lande and Schemske 1984; Charlesworth and Charlesworth 1987). However, for populations with extremely high inbreeding depression, such as some tree species and gynodioecious plants, it may be difficult to purge even the recessive lethals by close inbreeding, because if nearly all of the selfed offspring die before reproduction the population is then effectively outcrossed and no purging occurs unless the selfing rate exceeds a threshold value (Lande, Schemske, and Schultz 1994).

Substantial loss of fitness is an almost inevitable consequence of sudden reduction to very small population size, unless the population rapidly recovers to a large size thereby allowing selection to reverse the short-term effects of inbreeding and random genetic drift (e.g., Keller *et al.* 1994). The more gradual the reduction in population size, the greater the opportunity for purging recessive lethal mutations and avoiding a large part of the inbreeding depression. Thus, inbreeding depression is not simply proportional to the inbreeding coefficient routinely calculated for selectively neutral genes. The rule suggested by Franklin (1980) and Soulé (1980), supported by extensive practical experience in animal and plant breeding, is that inbreeding depression can be largely avoided in populations with effective sizes larger than $N_e = 50$. However, inbreeding depression may be more severe in natural environments than in laboratory populations (Jiménez *et al.* 1994) and more severe in stressful than in optimal environments (Keller *et al.* 1994; K. Biljsma pers. comm.).

Inbreeding depression can be readily reversed (at least temporarily) by introduction of several unrelated individuals into an inbred population, and permanent prevention of inbreeding depression can be accomplished by continued immigration every one or two generations of a single unrelated individual into each local population regardless of their size (Lande and Barrowclough 1987). Such a plan was recently instituted for the endangered Florida panther, motivated by strong circumstantial evidence of inbreeding depression in the small remnant population, and low genetic divergence from other conspecific populations (Hedrick 1995). Although this genetic augmentation may be necessary to reverse inbreeding effects (and not so high as to swamp any local adaptations), the Florida panther still faces the ecological threats of small population size due to past habitat destruction, and high mortality from automobile impacts.

Loss of genetic variation

Finite population size causes random changes in gene frequencies known as random genetic drift, attributable to Mendelian segregation and variation in family size, which results on average in a loss of genetic variance from a population. The expected proportion of selectively neutral genetic variance lost from a population per generation is $1/(2N_e)$, where N_e is the effective population size. For wild populations the effective size is usually substantially less than the actual size because of large variance in family size, unequal sex ratio among breeders, and fluctuations in population size through time (Wright 1969). Accounting for all these factors, the ratio of effective to actual size of wild populations is often on the order of 0.1 (Frankham 1995b). Weakly selected genes become effectively neutral with respect to the action of random genetic drift if the magnitude of selection on them is much less than $1/(2N_e)$ (Wright 1969).

To lose a large fraction of its genetic variance measured by heterozygosity in molecular genetic polymorphisms or heritable variance in quantitative characters, a population reduced to a small effective size must remain small for at least $2N_e$ generations. Following its loss in such a population "bottleneck," genetic variance can be replenished by immigration and/or mutation. An isolated population that passes through a

Genetic variance	Mutability	Time scale	Minimum N_e
Inbreeding depression	high	10^2	50
Quantitative characters	moderate	10^4	5000
Molecular heterozygosity			
point mutation	low	$2 \times 10^4 - 10^5$	$10^4 - 10^5$
microsatellite DNA	high	10^3	500

Table 1.2. Mutability, approximate time scale in generations for replenishment, and minimum effective population size (N_e) for maintaining typical levels, of different types of genetic variance in a randomly mating population. Molecular and quantitative variance are assumed to be quasi-neutral (excluding strongly selected mutations).

bottleneck must regain large size and remain large for a long time in order for mutation to restore normal levels of genetic variance. Metapopulation structure, with frequent local extinction and colonization, can reduce N_e of the metapopulation orders of magnitude below its actual size, mimicking the genetic effects of a population bottleneck (Wright 1940; Maruyama and Kimura 1980; Hedrick 1996).

Although all kinds of genetic variance are lost at the same rate by random genetic drift, in an isolated population different kinds of genetic variance are replenished at different rates depending on their mutability. Stable populations of different sizes also maintain unequal proportions of different kinds of genetic variance depending on the balance between random genetic drift, mutation and selection (see Table 1.2). Among small or moderate populations of given size, there may be substantial differences in the amount of inbreeding depression (Lacy *et al.* 1993), in the heritable variance in a given quantitative character (Bürger and Lande 1994), or in the heterozygosity at a given set of loci (Wright 1969). One should therefore not expect a close concordance in different types of genetic variance among populations of different size, contrary to the suggestion of Soulé (1980). In particular, populations with moderate effective size N_e on the order of 10^3 to 10^4 individuals, may maintain low molecular heterozygosity for point mutations, with substantial heritable variance in quantitative characters, and nearly normal inbreeding depression and heterozygosity for repeated DNA.

A low dispersal rate, on the order of a few individuals exchanged among populations per generation, is sufficient to prevent much genetic differentiation at quasineutral loci, such as most molecular genetic polymorphisms (Wright 1969; Crow and Kimura 1970). In contrast, adaptive differences among populations can be maintained by natural selection even under high levels of dispersal and gene flow (Endler 1977). Lack of differentiation between populations at molecular genetic loci therefore does not imply lack of adaptive differences. Thus, while molecular differentiation among populations is likely to imply adaptive divergence among populations, the converse is not true. It could therefore be a serious mistake to manage populations in different environments as a single unit, simply because no molecular differentiation among them has been detected, especially if morphological, behavioral, and physiological characteristics in which the populations might be adaptively differentiated have not been investigated.

To maintain typical levels of heritable variance in quantitative characters, based on experimental estimates of their mutability, Franklin (1980) and Soulé (1980) recommended a minimum effective population size of $N_e = 500$. Recent experimental evidence indicates that a large fraction of the mutational variance in quantitative characters is associated with recessive lethal and semi-lethal effects (Lopez and Lopez-Fanjul 1993a,b; Mackay, Lyman, and Jackson 1992), such that the quasi-neutral, potentially adaptive mutational variance is roughly one-tenth as large as previous estimates. Lande (1995) therefore suggested that the Franklin-Soulé number should be increased by a factor of 10, to $N_e = 5000$. Maintenance of rare alleles with major effects on disease resistance may require much larger populations (Roush and McKenzie 1987). However, populations that do not meet these simplistic criteria are not necessarily hopeless cases, for two reasons. First, if a population is sufficiently well adapted to its current environment, and if that does not change excessively, there may not be much need for adaptive evolution. Second, in a small population that recovers to large size, mutation can restore genetic variance and adaptability (Table 1.2). There are several examples of populations or species that have successfully recovered after being reduced to small numbers, such as the northern elephant seal (Hoelzel *et al.* 1993) and American bison (Miller 1990, pp. 38–39).

Quantitative (continuously varying) polygenic characters of morphology, behavior, and physiology generally are important for current adaptation, future adaptability, and population persistence. Quantitative characters typically are under stabilizing natural selection towards an intermediate optimum phenotype (that may fluctuate with time), such that extreme phenotypes are selected against. Like unconditionally deleterious mutations such as those contributing to inbreeding depression, heritable variance in quantitative characters therefore imposes a fitness cost or "genetic load" on a population, which is the price it must pay for future adaptability (Crow and Kimura 1970; Lande and Shannon 1996). Thus, under normal environmental conditions, including temporal fluctuations in the optimal phenotype, there is an optimal level of genetic variance for maintaining both current fitness and future adaptability.

Genetic variance in quantitative characters increases fitness and promotes population persistence primarily when environmental change is partially predictable; that is, under continued directional change, long-period high-amplitude cycles, or substantial autocorrelation in the optimum phenotype (Lande and Shannon 1996).

There is, however, a maximum rate of directional or random environmental change that a population can tolerate by adaptive evolution without becoming extinct, depending on the amount of genetic variability it can maintain (Lynch and Lande 1993; Bürger and Lynch 1995; Gomulkiewicz and Holt 1995; Lande and Shannon 1996). Rapid, extreme environmental changes, such as anthropogenic global warming, will place a premium on genetic variability and adaptability of many populations in fragmented environments during the coming centuries (see Conclusions).

Fixation of new mutations

In contrast to recessive lethal mutations that are generally kept at low frequencies by natural selection, random genetic drift can fix mildly deleterious mutations in a small population and gradually erode its fitness. Mildly deleterious mutations arise at many loci, with a total genomic rate on the order of one per generation in a variety of organisms; individually they produce an average fitness loss of a few to several percentages and are only partially dominant (nearly additive). When enough deleterious mutations become fixed, the population is genetically inviable ($r_{max} \leq 0$) and extinction rapidly ensues. For a population at carrying capacity in a constant environment, with no demographic stochasticity, the mean time until genetic inviability from fixation of new deleterious scales asymptotically as a power of the effective population size at carrying capacity; the power depends on the coefficient of variation of selection against new mutations (Lande 1994, 1995). For realistic distributions of selection on mildly deleterious mutations, the coefficient of variation is on the order of one (e.g., an exponential distribution of mutational effects on fitness), so the power is not very large (see Table 1, in Keightley 1994). With such a distribution of mutational effects it is the nearly neutral mutations, with selection coefficients close to $1/(2N_e)$, that do the most damage to the population, because strongly selected mutations rarely become fixed and more weakly selected mutations have relatively little impact on fitness (Lande 1994; Lynch, Conery, and Bürger 1995a,b).

With high initial fitness, even for extremely small populations it may take hundreds of generations for fixation of new mildly deleterious mutations to cause extinction. Advantageous, compensatory, and reverse mutations will completely prevent the erosion of fitness by deleterious mutations in sufficiently large populations. Thus, it is only for small populations with low fitness that the extinction risk from fixation of new deleterious mutations is a serious concern within the typical 100-year time scale of conservation planning. However, for populations of moderate size, with N_e up to few thousand, fixation of new mutations could substantially decrease their *long-term* viability (Lande 1995).

1.4 CONCLUSIONS

The primary anthropogenic causes of species declines produce a series of ecological and genetic effects that are finally expressed, and can be evaluated, in population dynamics and extinction risk. Land development causes habitat loss and fragmentation, which, along with overexploitation and artificial introductions of exotic species, causes population declines, creating small population effects and intensifying metapopulation dynamics. Demographic and genetic factors affecting small populations are involved in positive feedback loops with population decline, termed "extinction vortices" by Gilpin and Soulé (1986).

Management for recovery of an endangered species already reduced to small and/or fragmented populations requires consideration of all of the potential risk factors described above, as well as their interactions. Small-population effects are usually more a symptom than a cause of impending extinction, and treating them without addressing the underlying causes of population decline is not likely to prevent extinction (Lande 1988b; Caughley 1994). The scaling laws for mean time to extinction under different risks (Table 1.1) support the idea that deterministic population declines of anthropogenic origin are generally of much greater importance than stochastic factors as the main causes of species declines prior to their becoming endangered. This is especially important because it is often possible to ascertain the causes of deterministic declines and to reverse them through restoration and management actions (Caughley and Gunn 1996).

Habitat destruction and fragmentation restrict dispersal, and eliminate for many species what was the most important mechanism for population persistence in response to long-term climatic alterations: change of geographic distribution (Pease, Lande, and Bull 1989; Peters and Lovejoy 1992). In response to previous periods of global warming and cooling associated with glacial cycles, species often changed their geographic range while maintaining essentially the same phenotype except perhaps for changes in body size (e.g., Coope 1979; Smith, Betancourt, and Brown 1995). Species restricted to isolated habitat fragments and reserves must instead rely either on their limited physiological tolerances, or on evolutionary adaptation *in situ*, to survive rapid global warming in the coming centuries. A small proportion may be aided by accidental or deliberate artificial transport. Persistence of many species during the next millennium will therefore come increasingly to depend on maintaining ample genetic variation for adaptive evolution, and on having natural or artificial opportunities for dispersal.

Despite heightened public awareness and concern about environmental issues, national and international efforts at conservation and restoration remain largely inadequate, as most nations promote continued human population growth, more land development, increased resource exploitation, and anthropogenic global warming. Although politicians and societies rarely make plans on time scales longer than decades, conservation biologists must increasingly plan on time scales of centuries, millennia, and longer, if our attempts to preserve some fraction of existing biodiversity are to have any lasting effect.

ACKNOWLEDGMENTS

This work was supported by NSF grant DEB-9225127.

REFERENCES

Allee, W. C., Emerson, A. E., Park, O., Park, T., and Schmidt, K. P. (1949). *Principles of Animal Ecology*. Philadelphia: Saunders.

Allendorf, F. W., and Waples, R. S. (1996). Conservation and genetics of salmonid fishes. In Avise, J. C. and Hamrick, J. L. (eds.), *Conservation genetics: Case histories from nature*, 238–280, New York: Chapman and Hall.

Andreassen, H. P., Halle, S., and Ims, R. A. (1996). Optimal width of movement corridors for root voles: Not too narrow and not too wide. *J. Appl. Ecol.*, *33*, 63–70.

Andrewartha, H. G., and Birch, L. C. (1954). *The distribution and abundance of animals*. Chicago: University of Chicago Press.

Arnold, S. J. (1995). Monitoring quantitative genetic variation and evolution in captive populations. In Ballou, J., Gilpin, M and Foose, T. J. (eds.), *Population management for survival and recovery: analytical methods and strategies in small populations,* 295–317, New York: Columbia University Press.

Atkinson, I. (1989). Introduced animals and extinctions. In Western, D. and Pearl, M. (eds.), *Conservation for the twenty-first century*, 59–75, Oxford: Oxford University Press.

Beddington, J. R., and May, R. M. (1977). Harvesting populations in a randomly fluctuating environment. *Science*, *197*, 463–465.

Brown, J. H., and Kodric-Brown, A. (1977). Turnover rates in insular biogeography: Effect of immigration on extinction. *Ecology*, *58*, 445–449.

Bürger, R., and Lande, R. (1994). On the distribution of the mean and variance of a quantitative trait under mutation-selection-drift balance. *Genetics, 138*, 901–912.

Bürger, R., and Lynch, M. (1995). Evolution and extinction in a changing environment: A quantitative-genetic analysis. *Evolution, 49*, 151–163.

Burkey, T. V. (1989). Extinction in nature reserves: The effect of fragmentation and the importance of migration between reserve fragments. *Oikos, 55*, 75–81.

Burkey, T. V. (1995). Extinction rates in archipelagoes: Implications for populations in fragmented habitats. *Conserv. Biol., 9*, 527–541.

Caughley, G. (1994). Directions in conservation biology. *J. Anim. Ecol., 63*, 215–244.

Caughley, G., and Gunn, A. (1996). *Conservation biology in theory and practice.* London: Blackwell Science.

Charlesworth, D., and Charlesworth, B. (1987). Inbreeding depression and its evolutionary consequences. *Ann. Rev. Ecol. Syst., 18*, 237–268.

Clark, C. W. (1973). The economics of overexploitation. *Science, 181*, 630–634.

Clark, C. W. (1990). *Mathematical bioeconomics, 2d ed.,* New York: Wiley.

Coope, G. R. (1979). Late Cenozoic fossil Coleoptera: Evolution, biogeography, and ecology. *Ann. Rev. Ecol. Syst., 10*, 247–267.

Crow, J. F., and Kimura, M. (1970). *Introduction to population genetics theory.* New York: Harper and Row.

DeMauro, M. M. (1993). Relationship of breeding system to rarity in the Lakeside Daisy (*Hymenoxys acaulis var. glabra*). *Conserv. Biol., 7*, 542–550.

Doak, D. (1989). Spotted owls and old growth logging in the Pacific Northwest. *Conserv. Biol., 3*, 389–396.

Dobson, A. P., and May, R. M. (1986). Disease and conservation. In Soulé, M. E. (ed.), *Conservation biology: The science of scarcity and diversity,* 345–365, Sunderland, Mass.: Sinauer.

Dobzhansky, T. (1970). *Genetics of the evolutionary process.* New York: Columbia University Press.

Downes, S. J., Handasyde, K. A., and Elgar, M. A. (1997). The use of corridors by mammals in fragmented Australian eucalypt forests. *Conserv. Biol., 11*, 718–726.

Endler, J. (1977). *Geographic variation, speciation, and clines.* Princeton: Princeton University Press.

Falconer, D. S., and Mackay, T. F. C. (1996). *Introduction to Quantitative Genetics, 4th ed.,* London: Longman.

Frankham, R. (1995a). Inbreeding and extinction: A threshold effect. *Conserv. Biol., 9*, 792–799.

Frankham, R. (1995b). Effective population size/adult population size ratios in wildlife: A review. *Genet. Res., Camb., 66*, 95–107.

Franklin, I. R. (1980). Evolutionary change in small populations. In Soulé, M. E., and Wilcox, B. A. (eds.), *Conservation biology: An evolutionary-ecological perspective.* 135–149, Sunderland, Mass.: Sinauer.

Gilpin, M. E., and Soulé, M. E. (1986). Minimum viable populations: Processes of species extinction. In Soulé, M. E. (ed.), *Conservation biology: The science of scarcity and diversity.* 19–34, Sunderland, Mass.: Sinauer.

Gomulkiewicz, R., and Holt, R. D. (1995). When does evolution by natural selection prevent extinction? *Evolution, 49*, 201–207.

Grant, P. R. and Price, T. D. (1981). Population variation in continuously varying traits as an ecological genetics problem. *Am. Zool., 21*, 795–811.

Groombridge, B., ed. (1992). *Global biodiversity: Status of the earth's living resources.* London: Chapman and Hall.

Hanski, I., and Gilpin, M. E., eds. (1997). *Metapopulation Biology*. London: Academic Press.

Hanski, I., and Gyllenberg, M. (1993). Two general metapopulation models and the core-satellite species hypothesis. *Am. Nat.*, *142*, 17–41.

Hanski, I., Poyry, J., Pakkala, T., and Kuussaari, M. (1995). Multiple equilibria in metapopulation dynamics. *Nature*, *377*, 618–621.

Hedrick, P. W. (1995). Gene flow and genetic restoration: The Florida panther as a case study. *Conserv. Biol.*, *9*, 996–1007.

Hedrick, P. W. (1996). Bottleneck(s) or metapopulation in cheetahs. *Conserv. Biol.*, *10*, 897–899.

Hess, G. (1996). Disease in metapopulation models: Implications for conservation. *Ecology*, *77*, 1617–1632.

Hoelzel, A. R., Halley, J., O'Brien, S. J., Campagna, C., Arnbom, T., LeBoeuf, B., Ralls, K., and Dover, G. A. (1993). Elephant seal genetic variation and the use of simulation models to investigate historical population bottlenecks. *J. Hered.*, *84*, 443–449.

Jiménez, J. A., Hughes, K. A., Alaks, G., Graham, L., and Lacy, R. C. (1994). An experimental study of inbreeding depression in a natural habitat. *Science*, *266*, 271–273.

Keightley, P. D. (1994). The distribution of mutation effects on viability in *Drosophila melanogaster*. *Genetics*, *138*, 1315–1322.

Keller, L. F., Arcese, P., Smith, J.N.M., Hochachka, W. M., and Stearns, S. C. (1994). Selection against inbred song sparrows during a natural population bottleneck. *Nature*, *372*, 356–357.

Kierstead, H., and Slobodkin, L. B. (1953). The sizes of water masses containing plankton bloom. *J. Mar. Res.*, *12*, 141–147.

Lacy, R. C., Petric A., and Warneke, M. (1993). Inbreeding and outbreeding in captive populations of wild animal species, In Thornhill, N. W. (ed.), *The natural history of inbreeding and outbreeding: Theoretical and empirical perspectives*, 352–374, Chicago: University of Chicago Press.

Lande, R. (1987). Extinction thresholds in demographic models of territorial populations. *Am. Nat.*, *130*, 624–635.

Lande, R. (1988a). Demographic models of the northern spotted owl (*Strix occidentalis caurina*). *Oecologia*, *75*, 601–607.

Lande, R. (1988b). Genetics and demography in biological conservation. *Science*, *241*, 1455–1460.

Lande, R. (1993). Risks of population extinction from demographic and environmental stochasticity and random catastrophes. *Am. Nat.*, *142*, 911–927.

Lande, R. (1994). Risk of population extinction from fixation of new deleterious mutations. *Evolution*, *48*, 1460–1469.

Lande, R. (1995). Mutation and conservation. *Conserv. Biol.*, *9*, 782–791.

Lande, R. (1998). Demographic stochasticity and Allee effect on a scale with isotropic noise. *Oikos*, *83(2)*, 353-358.

Lande, R., and Barrowclough, G. F. (1987). Effective population size, genetic variation, and their use in population management, In Soulé, M. E. (ed.), *Viable populations for conservation*, 87–123, Cambridge: Cambridge University Press.

Lande, R., Engen, S., and Sæther, B.-E. (1994). Optimal harvesting, economic discounting, and extinction risk in fluctuating populations. *Nature, Lond.*, *372*, 88–90.

Lande, R., Engen, S., and Sæther, B.-E. (1995). Optimal harvesting of fluctuating populations with a risk of extinction. *Am. Nat.*, *145*, 728–745.

Lande, R., Engen, S., and Sæther, B.-E. (1998). Extinction times in finite metapopulation models with explicit local dynamics. *Oikos, 83(2)*, 383-389.

Lande, R., and Schemske, D. W. (1984). The evolution of self-fertilization and inbreeding depression in plants. I. Genetic models. *Evolution, 39*, 24–40.

Lande, R., Schemske, D. W., and Schultz, S. T. (1994). High inbreeding depression, selective interference among loci, and the threshold selfing rate for purging recessive lethal mutations. *Evolution, 48*, 965–978.

Lande, R., and Shannon, S. (1996). The role of genetic variability in adaptation and population persistence in a changing environment. *Evolution, 50*, 434–437.

Lande, R., Sæther, B.-E., and Engen, S. (1997). Threshold harvesting for sustainability of fluctuating resources. *Ecology, 78*, 1341–1350.

Levin, D. A., Francisco-Ortega, J., and Jansen, R. K. (1996). Hybridization and the extinction of rare plant species. *Conserv. Biol., 10*, 10–16.

Levins, R. (1970). Extinction. In Gerstenhaber, M. (ed.), *Some mathematical problems in biology,* 77–107, Providence, RI: American Mathematical Society.

Lewontin, R. C., and Birch, L. C. (1966). Hybridization as a source of variation for adaptation to new environments. *Evolution, 20*, 315–336.

Lopez, M. A., and Lopez-Fanjul, C. (1993a). Spontaneous mutation for a quantitative trait in *Drosophila melanogaster*. I. Response to artificial selection. *Genet. Res., Camb., 61*, 107–116.

Lopez, M. A., and Lopez-Fanjul, C. (1993b). Spontaneous mutation for a quantitative trait in *Drosophila melanogaster*. II. Distribution of mutant effects on the trait and fitness. *Genet. Res., Camb., 61*, 117–126.

Lovejoy, T. E., et. al. (1986). Edge and other effects of isolation on Amazon forest fragments. In Soulé, M. E. (ed.), *Conservation biology: The science of scarcity and diversity.* 257–285, Sunderland, Mass.: Sinauer.

Ludwig, D., Hilborn, R., and Walters, C. (1993). Uncertainty, resource exploitation, and conservation: Lessons from history. *Science, 260*, 17, 36.

Lynch, M., Conery, J., and Bürger, R. (1995a). Mutational meltdown in sexual populations. *Evolution, 49*, 1067–1080.

Lynch, M., Conery, J., and Bürger, R. (1995b). Mutation accumulation and the extinction of small populations. *Am. Nat., 146*, 489–518.

Lynch, M., and Lande, R. (1993). Evolution and extinction in response to environmental change. In Karieva, P., Huey, R., and Kingsolver, J. (eds.) *Biotic interactions and global change,* 234–250, Sunderland, Mass.: Sinauer.

MacArthur, R. H., and Wilson, E. O. (1967). *The theory of island biogeography.* Princeton: Princeton University Press.

Mackay, T. F. C., Lyman, R. F., and Jackson, M. S. (1992). Effects of *P* element insertion on quantitative traits in *Drosophila melanogaster*. *Genetics, 130*, 315–332.

Maruyama, T., and Kimura, M. (1980). Genetic variation and effective population size when local extinction and recolonization of subpopulations are frequent. *Proc. Natl. Acad. Sci. USA, 77*, 6710–6714.

May, R. M. (1976). Harvesting whale and fish populations. *Nature, Lond., 263*, 91–92.

May, R. M., Beddington, J. R., Horwood, J. W., and Shepherd, J. G. (1978). Exploiting natural populations in an uncertain world. *Math. Bio., 42*, 219–252.

McKelvey, K., Noon, B. R., and Lamberson, R. H. (1993). Conservation planning for species occupying fragmented landscapes: The case of the northern spotted owl. In Karieva, P., Huey, R., and Kingsolver, J. (eds.), *Biotic interactions and global change,* 424–450, Sunderland, Mass.: Sinauer.

Miller, G. T., Jr. (1990). *Living in the Environment, 6th ed.,* Belmont, CA: Wadsworth.

Myers, R. A., Bridson, J., and Barrowman, N. J. (1995). Summary of worldwide spawner and recruitment data. *Canadian Technical Report of Fisheries and Aquatic Sciences,* 2024.

Nehlsen, W., Williams, J. E., and Lichatowich, J. A. (1991). Pacific salmon at the crossroads: Stocks at risk from California, Oregon, Idaho, and Washington. *Fisheries, 16,* 4–21.

ODEC (1991). *The State of the Environment.* Paris: Organization for Economic Cooperation and Development.

Okubo, A. (1980). *Diffusion and ecological problems: Mathematical models.* Berlin: Springer-Verlag.

Pease, C.M., Lande, R., and Bull, J. J. (1989). A model of population growth, dispersal and evolution in a changing environment. *Ecology, 70,* 1657–1664.

Peters, R. L., and Lovejoy, T. E. (1992). *Global warming and biological diversity.* New Haven: Yale University Press.

Pimm, S. L. (1991). *The balance of nature?* Chicago: University of Chicago Press.

Ralls, K., and Ballou, J. D. (1983). Extinction: Lessons from zoos. In Schonewald-Cox, C. M., Chambers, S. M., MacBryde, B., and Thomas, W. L. (eds.), *Genetics and conservation: A reference for managing wild animal and plant populations,* 164–184, Menlo Park, CA: Benjamin-Cummings.

Ratner, S., Lande, R., and Roper, B. B. (1997). Population viability analysis of spring chinook salmon in the South Umpqua river, Oregon. *Conserv. Biol., 11,* 879–889.

Robinson, S. K., Thompson, F. R. III, Donovan, T. M., Whitehead, D. R., and Faaborg, J. (1995). Regional forest fragmentation and the nesting success of migratory birds. *Science, 267,* 1987–1990.

Roush, R. T., and McKenzie, J. A. (1987). Ecological genetics of insecticide and acaricide resistance. *Ann. Rev. Entomol., 32,* 361–380.

Redford, K. H. (1992). The empty forest. *Bioscience, 42,* 412–422.

Richter-Dyn, N., and Goel, N. S. (1972). On the extinction of a colonizing species. *Theoret. Popul. Biol., 3,* 406–433.

Rosenberg, A. A., Fogarty, M. J., Sissenwine, M. P., Beddington, J. R. and Shepherd, J. G. (1993). Achieving sustainable use of renewable resources. *Science, 262,* 828–829.

Seehausen, O., van Alphen, J.J.M., and Witte, F. (1997). Cichlid fish diversity threatened by eutrophication that curbs sexual selection. *Science, 277,* 1808–1811.

Simmons, M. J., and Crow, J. F. (1977). Mutations affecting fitness in *Drosophila* populations. *Ann. Rev. Genet., 11,* 49–78.

Smith, F. A., Betancourt, J. L., and Brown, J. H. (1995). Evolution of body size in the woodrat over the past 25,000 years of climate change. *Science, 270,* 2012–2014.

Soulé, M. E. (1980). Thresholds for survival: Maintaining fitness and evolutionary potential. In Soulé, M. E., and Wilcox, B. A. (eds.), *Conservation biology: An evolutionary-ecological perspective,* 151–169, Sunderland, Mass.: Sinauer.

Stokes, T. K., McGlade, J. M., and Law, R. (1993). The exploitation of evolving resources. In *Lecture Notes in Biomathematics,* vol. 99. Berlin: Springer-Verlag.

Thomas, J. W., Forsman, E. D., Lint, J. B., Meslow, E. C., Noon, B. R., and Verner, J. (1990). A conservation strategy for the northern spotted owl. Washington, D.C.: U.S. Government Printing Office.

Vitousek, P. M. (1988). Diversity and biological invasions of oceanic islands. In Wilson, E. O. (ed.), *Biodiversity,* 181–189, Washington, D.C.: National Academy Press.

Wayne, R. K. (1996). Conservation genetics in the Canidae. In Avise, J. C., and Hamrick, J. L. (eds.), *Conservation genetics: Case histories from nature*, 75–118. New York: Chapman and Hall.

Wright, S. (1940). Breeding structure of populations in relation to speciation. *Am. Nat.*, *74*, 232–248.

Wright, S. (1969). *Genetics and the evolution of populations.* Vol. 2. *The theory of gene frequencies.* Chicago: University of Chicago Press.

Young, T. P. (1994). Natural die-offs of large mammals: Implications for conservation. *Conserv. Biol.*, *8*, 410–418.

2

Genes, Demes, and Plant Conservation

KENT E. HOLSINGER, ROBERTA J. MASON-GAMER,
AND JEANNETTE WHITTON

SUMMARY. Although conservation biologists have often expressed concern about loss of genetic diversity in endangered plants, loss of diversity is more likely to be a symptom of endangerment than its cause. Changes in the genetic structure of a plant population are likely to threaten its persistence only if they involve loss of self-incompatibility alleles or genetic assimilation through hybridization with a reproductively compatible relative. This chapter will demonstrate how an understanding of the patterns of genetic variation within and among populations may help conservation biologists to identify evolutionarily distinct populations worthy of conservation concern. These factors may provide insight into the extent of demographic connections among existing populations. One must interpret data from molecular markers with caution, however, because patterns of variation for those markers may be quite different from those for gentoypes at polygenic traits likely to be involved in future adaptive responses.

INTRODUCTION

Of the roughly 250,000 plant species known to have been extant in historical times, almost 1000 have become extinct in the last century, and more than sixty times that many may become extinct in the next fifty years (Raven 1987). The great bulk of these species are found in tropical latitudes, especially in lowland tropical forests and elsewhere, but the number at risk in temperate latitudes is by no means small. The 1997 International Union for Conservation of Nature and Natural Resources's (IUCN) *Red List of Threatened Plants*, for example, lists almost 30% of species occuring in the United States as threatened (Walter and Gillett 1998). Even if you exclude the 500 species listed from the state of Hawaii, the number of plant species at risk of extinction exceeds that of all bird, mammal, reptile, and amphibian species that breed in Canada and the United States. Moreover, the same pattern is repeated on much smaller scales. In Connecticut, nearly 300 plant species are regarded as endangered, threatened, or of special concern within the state. Of those, nearly one-half are no longer found in the state, although most still occur elsewhere. None of the 60 species of mammals and only 14 of the 114 species of birds that breed in the state are threatened within its borders (Holsinger 1995a).

As Lande (Chapter 1 in this volume) has emphasized, the primary causes of species extinctions today are deterministic, anthropogenic effects, especially those associated with habitat loss, habitat destruction, or overexploitation. In that context, the contributions that genetic concepts and genetic techniques can make to arrest the extinction crisis are necessarily somewhat limited. The first task of any endangered species management program must be to reverse the deterministic threats to persistence. In addition, the number of plant species at risk is so great that only a small fraction of those that are endangered can receive individual management attention. There are, nevertheless, some ways in which genetics may contribute to plant conservation, and we attempt to outline them in this chapter.

2.1 Genetic Diversity and Population Viability

Genetic drift and the loss of genetic diversity associated with it is the primary genetic concept that conservation biologists have adopted in assessing genetic extinction risks. Clearly, loss of genetic diversity threatens the ability of populations to respond adaptively to future environmental change, and it may even threaten the short-term persistence of populations if the expressed effects of inbreeding depression are sufficiently severe. Lande (1988, and Chapter 1 in this volume) has argued convincingly, however, that populations large enough to buffer the risks of environmental stochasticity are likely to be far larger than necessary to buffer risks associated with loss of genetic diversity. Even the population sizes necessary to allow for an adaptive response to future environmental change are likely to be far smaller than those necessary to protect populations against ecologically driven extinctions in the face of typical levels of environmental variability. Thus, levels of genetic diversity are likely to be a poor indicator of population viability.

There are in addition several reasons to doubt that loss of genetic diversity will often cause populations to become endangered, although lack of genetic diversity may increase the threat to populations that are already endangered (Holsinger 1995b; Holsinger and Vitt 1997). First, low-frequency alleles have little to do with any immediate response to natural selection. Fisher's fundamental theorem of natural selection

| Allele | Effective population size | |
frequency	$N_e = 100$	$N_e = 1000$
0.001	3	30
0.01	19	186
0.05	63	630

Table 2.1. This table shows the mean number of generations until a rare allele is lost to genetic drift.

shows that the response of any population to natural selection is directly proportional to the additive genetic variance in fitness (see, for example, Crow and Kimura 1970, and see Frank and Slatkin 1992 for a recent discussion of its proper interpretation). Alleles occurring in moderate frequencies contribute most to determining the amount of additive genetic variation present, because genotypic differences among individuals within a population are accounted for primarily by differences in those alleles. As a result, rare alleles contribute very little to variation in fitness among individuals. Thus, those alleles most likely to be lost as a result of genetic drift — rare alleles — are also the least likely to contribute to any immediate adaptive response to natural selection.

Second, alleles that are currently in low frequency are not likely to be an important component of adaptive responses to future environmental change, because they are likely to be lost before such changes occur. Rare alleles, which are the ones most likely to be lost as a result of genetic drift, have a short expected lifetime unless natural selection is acting to maintain them in populations. Only

$$\bar{t} = -4N_e \left(\frac{p}{1-p}\right) \ln p$$

generations elapse, on average, before a neutral allele currently present at frequency p in the population is lost as a result of genetic drift (Crow and Kimura 1970; Ewens 1979). A neutral allele present in a frequency of 0.1%, for example, will be lost in fewer than thirty generations in a population with an effective size of 1,000, and one present in a frequency of 1% will be lost in fewer than twenty generations in a population with an effective size of 100 (Table 2.1). An allele that is selectively disfavored will be lost even faster.

These considerations make it clear that any alleles currently present in low frequency are unlikely to be present more than a few tens of generations from now. Adaptation to future environmental changes, therefore, is likely to result from genetic differences among individuals that result either from allelic differences for alleles that are currently present in relatively high frequency or from allelic differences introduced through mutation.[1] It is unlikely to result from an increase in frequency of alleles currently found in low frequency. One exception to this general prediction might occur if natural selection acts to maintain rare alleles, for example, frequency-dependent selection favoring rare self-incompatibility alleles. If natural selection is maintaining these rare alleles, however, it is unlikely that we need to worry about losing them as a result of genetic drift, except in very small populations, because natural selection itself will act to maintain them.

Third, environmental variability generally poses a much greater threat to the persistence of populations than immediate fitness impacts associated with accumulation of deleterious mutations in small populations. In large populations unfavorable mutations are very unlikely to be fixed. In small populations, however, there is an appreciable chance that disfavored mutations at any locus will be fixed as a result of genetic drift. Once fixed, the maximum fitness attainable by any individual in that population is reduced, and the reproductive capacity of the entire population may be substantially lowered. The accumulation of these disfavored mutations is analogous to the accumulation of mutations in asexual lines, that is, Muller's ratchet (Muller 1964). As several authors (Lynch and Lande 1993; Lande 1994; Bürger and Lynch 1995; Lande 1995; Lynch *et al.* 1995; Lande and Shannon 1996) have shown, this effect is greater than many previously believed possible, and it is largest for slighty deleterious mutations.

Deleterious mutations of large effect are unlikely to be fixed as a result of drift, even in very small populations. Deleterious mutations of very mild effect are almost as likely to be fixed as neutral mutations, but they have very little impact on the potential maximum fitness of individuals. Slightly deleterious mutations have both an appreciable probability of fixation and an appreciable impact on the potential maximum fitness of individuals in the affected population. Even when most newly arising mutations are only slightly deleterious, however, populations larger than a few hundred are expected to persist for tens or hundreds of thousands of generations if environmental stochasticity is ignored (Table 2.2: mutation accumulation). When the variance in population growth rate exceeds the mean intrinsic rate of increase, however, populations as large as 10,000 are expected to persist for only a few hundred generations (Table 2.2: periodic catastrophes). These results strongly suggest that mutation accumulation is less likely to pose a threat to the continued persistence of most populations than typical levels of environmental variation.

---

[1]See also the discussion below on indirect genetic threats to persistence.

Population size	Mean number of generations to extinction	
	Mutation accumulation	Periodic catastrophe
10	99	111
100	2149	280
500	43878	406
1000	170911	461
10000	1.7×10^7	645

Table 2.2. The table shows the mean number of generations until extinction occurs for a given population size under two scenarios. For both scenarios the mean intrinsic rate of increase is 1% per generation. In the mutation accumulation scenario, one deleterious mutation occurs, on average, in each generation; the mean selection coefficient agains deleterious alleles is 0.025, and the coefficient of variation in the selection coefficient is 1. In the periodic catastrophe scenario, the time between catastrophes is exponentially distributed with a mean of approximately 67 generations, and 75% of the population is eliminated in each catastrophe. The mutation accumulation results are based upon Lande (1994); the periodic catastrophe results are based upon Ewens *et al.* (1987).

Taken together, these considerations suggest that loss of genetic diversity and many other features associated with life in small populations are more a symptom of endangerment than their cause (Caughley 1994; Holsinger and Vitt 1997). If a particular rare plant lacks genetic diversity, for example, that lack of diversity is more likely to be a consequence of the process that made it rare in the first place than it is to be the cause of the decline. Moreover, those species that have always been rare have demonstrated by the very fact of their continued existence their ability to cope with the problems of life in small populations, be they demographic or genetic (Holsinger and Gottlieb 1991). Indeed, recent theoretical work (Kirkpatrick and Barton 1997) suggests that the level of genetic variability has little effect on persistence or range size of species with small populations. Range size appears to be related to the level of genetic variability in populations only in species that are already quite abundant.

2.2 GENETIC THREATS TO PERSISTENCE

It is tempting to conclude from this set of observations that changes in the genetic structure of plant populations will never pose a threat to their persistence; to conclude, for example, that "it's the demography, stupid." Tempting, but wrong. There are two important cases where changes in the genetic structure of plant populations may threaten their persistence:

1. Loss of self-incompatibility alleles may pose a direct threat to the reproductive capacity of individuals in those plant species with a genetically determined self-incompatibility system.
2. Hybridization of rare species with widespread relatives with which they are reproductively compatible may lead to extinction of the rare species through genetic assimilation.

In addition, loss of genetic diversity associated with the loss of locally adapted populations may pose an indirect threat to persistence of some species.

Loss of Self-incompatibility Alleles

Loss of self-incompatibility poses a direct, immediate threat to the reproductive success of the rare Lakeside Daisy (*Hymenoxys acaulis*). The Illinois population of this species, which has been listed as an endangered species by the United States Fish and Wildlife Service, had failed to set seed for more than fifteen years prior to an analysis of its reproductive biology (DeMauro 1993). A series of controlled crosses, including plants from Ohio, revealed that all members of the Illinois population belong to a single compatibility type, meaning than no compatible pollinations within the Illinois population were possible. Even the Ohio populations, all of which had more than one compatibility group present, mostly had only three or four compatibility groups present. Only one population had as many as nine. As a result, most possible outcrosses produce less than 2% seed set. The small number of compatibility groups within the Ohio populations limits the reproductive success of individuals within them and threatens the long-term persistence of those populations. Furthermore, to reestablish a long-term, self-sustaining population in Illinois, it was necessary to introduce genotypes from Ohio populations to increase the number of compatibility groups within the the Illinois population.

Genetic Assimilation

Hybridization poses a threat to rare species quite different from the direct reproductive impact that loss of self-incompatibility alleles poses. If a rare species hybridizes frequently with a reproductively compatible common one, the rare one may be hybridized out of existence (Cade

Taxon/Individual	RAPD marker						6PGD
Cercocarpus betuloides var. *blancheae*	+	+	−	−	+	−	aa
Cercocapus traskiae	−	−	+	+	−	+	bb
Hybrids							
BL	−	−	+	+	−	−	aa
E	−	+	−	−	−	−	bb
F	+	+	−	−	+	−	ab
G	+	+	+	+	−	−	ab
WB3	−	−	−	−	−	+	aa

Table 2.3. Hybridization in mountain mahoganies. (Modified from Rieseberg and Gerber 1995.)

1983; Rieseberg 1991; Ellstrand 1992; Ellstrand and Elam 1993). Although its distinctive alleles may be assimilated into the common species, the rare species itself may cease to exist. One particularly striking example comes from Catalina Island off the coast of southern California, where the last remaining population of the rare Catalina Island mountain mahogany (*Cercocarpus traskiae*) has been reduced to only eleven trees. For a variety of reasons, previous workers had suspected that some of these trees might be hybrids produced through crossing with a widespread species (*C. betuloides* var. *blancheae*) found both throughout Catalina Island and on the mainland. To investigate this possibility Rieseberg and Gerber (1995) assayed variation in markers detected through the use of randomly amplified polymorphic DNA segments (RAPDs) and allozyme electrophoresis. Each species has a series of diagnostic markers that distinguish it from its congener. The data summarized in Table 2.3 makes it clear that five of the eleven trees found on the island combine genotypes of the two distinct species, indicating that those five trees are of hybrid origin. Furthermore, analysis of several seedlings that were collected and analyzed in the vicinity of the parental trees showed diagnostic markers characteristic of both parents, suggesting that much of the reproduction occurring in this population is occurring through hybridization with a widespread relative. In short, we have a situation where almost half of the remaining trees appear to be hybrid, many of the seedlings that are being produced are hybrid, and the most immediate threat to the rare species is not the result of a declining number of individuals but through genetic assimilation.[2]

[2]Obviously, the very small number of individuals remaining makes the threat of genetic assimilation greater than if the remaining population were larger.

Indirect Genetic Threats to Persistence

Although loss of genetic diversity within populations is unlikely to pose a direct threat to their persistence, loss of genetic diversity associated with loss of entire populations could pose a long-term threat to the persistence of species. Many studies have shown that plant populations are ecologically differentiated from one another, often on surprisingly small scales (see, for example, Aitken and Libby 1994; Couland and McNeilly 1992; Schmitt and Gamble 1990; Stanton *et al.* 1997; Stratton 1994). As a result, the fitness of transplants moved to a environment similar to that from which they were collected is generally less than half that of residents (Bradshaw 1984). Thus, it may be difficult to reestablish populations once they are lost. Not only does a continuing loss of populations expose the species to ever greater demographic risks, but the difficulty of reestablishing populations with nonresident propagules also suggests that attempts to reverse the loss of populatons will become more and more difficult as more and more populations are lost. Loss of genetic diversity represented by differences among populations, therefore, may serve to accelerate a demographically driven decline.

2.3 INTERPRETING PATTERNS OF GENETIC DIVERSITY

Genetic threats to persistence appear to be less pervasive than was widely believed a decade ago. As a result, the most important contributions geneticists can now make to endangered species conservation are likely to result from applying techniques they have developed to address questions in systematics, evolution, and ecology. In particular, molecular techniques now provide powerful techniques for identifying populations that are evolutionarily independent of one another, for unraveling the pattern of historical connections among geographically disjunct populations, and for detecting demographic processes that may be difficult or impossible to discover in any other way (see, for example, the discussion in Avise 1994; Avise 1995; Holsinger and Vitt 1997). To illustrate the potential of such markers for plant conservation genetics, we describe how the analysis of chloroplast DNA (cpDNA) restriction-site polymorphisms has helped to clarify evolutionary and demographic questions in two different genera of the sunflower family, *Coreopsis* and *Crepis*.

Restriction Site Variation in cpDNA

Surveys of cpDNA diversity within plant populations have generally suggested that little diversity would be found (see, for example, Banks and

Birky (1985), but see Soltis *et al.* (1992) for other surveys revealing variation within species). Within *Coreopsis*, in particular, an earlier study of the nine members of section (sect.) *Coreopsis* found only four haplotypes in the entire section (Crawford *et al.* 1990). These results would suggest that there is little to be learned from restriction-site analyses of cpDNA at the population level. Many earlier studies, however, assayed for variation with restriction enzymes that cut the chloroplast genome relatively infrequently. Thus, the lack of within-species diversity found should not be surprising.

To enhance the possibility of detecting significant cpDNA restriction-site variation within and among populations, the studies discussed below used the following general protocol:

> Leaf samples were collected either from individuals (the *Coreopsis* studies) or from groups of individuals belonging to a well-defined morphological type (the *Crepis* study). DNA was extracted from these samples according to standard procedures (Doyle and Doyle 1987), and samples were digested separately with six or more frequent-cutting restriction enzymes. Restriction fragments were separated on 1.25%–1.5% agarose gels and bidirectionally blotted to nylon membranes. Membrane-bound DNA fragments were hybridized with ^{32}P-labeled, cloned fragments of the lettuce chloroplast genome (Jansen and Palmer 1987), corresponding to the most variable half of the genome. (Kim *et al.* 1992)

As a result of focusing on a portion of the chloroplast genome already known to be the most variable, and of assaying variation with restriction enzymes that cut the chloroplast genome frequently, each study found substantial cpDNA restriction-site variation within and among populations.

The Origin of *Coreopsis nuecensis*

Coreopsis nuecensoides and *C. nuecensis* are a pair of closely related, narrowly distributed species, endemic to southeastern Texas. Although their ranges overlap, they are not known to occur in mixed populations (Smith 1974). They are distinguished from one another primarily by the presence (*C. nuecensis*) or absence (*C. nuecensoides*) of hairs on the involucral bracts, although *C. nuecensoides* is also weakly perennial in the greenhouse while *C. nuecensis* is strictly annual. A long series of analyses considering variation in chromosome structure, chromosome number, and allozymes is consistent with the hypothesis that *C. nuecensis* is recently derived from within *C. nuecensoides* (Smith 1974; Crawford and Smith 1982; Crawford and Smith 1984; Cosner and Crawford

1990). To further investigate this possibility Mason-Gamer *et al.* (1999) examined individual-level patterns of restriction-site variation in samples collected from several populations of each species.

Thirteen distinct haplotypes were identified in the 114 individuals sampled in this study, and the pattern of variation discovered is broadly consistent with a population bottleneck during the divergence of *Coreopsis nuecensis* from *C. nuecensoides* (Fig. 2.1). Seven haplotypes are restricted to *C. nuecensoides*, the presumed progenitor, five to *C. nuecensis*, the presumed derivative, and one is shared between them. Phylogenetic analysis of the restriction-site differences is completely consistent with the hypothesized progenitor-derivative relationship. Notice, however, that most of the diversity in both species appears to have originated after they diverged from one another. As a result, it is impossible to assess the extent of any bottleneck associated with the speciation event by using these data. Nonetheless, this study illustrates that chloroplast DNA markers can provide insight into patterns of evolutionary relatedness and the extent of evolutionary distinctness at the level of population or species divergence in plants. The next study illustrates that it may even be possible to gain a sense of the historical relationships among populations within a single species.

Genetic Diversity in *Coreopsis grandiflora*

Another member of *Coreopsis* sect. *Coreopsis* poses a quite different evolutionary problem. *Coreopsis grandiflora* is distributed in two major population groups. One center of distribution is found in northwestern Georgia, the other in northeastern Oklahoma and Arkansas; relatively few populations are found in between. Nonetheless, two of the same taxonomically recognized varieties occur in both of these distribution centers. Thus, an important evolutionary question is to determine whether those varieties are an example of morphological convergence or a reflection of evolutionary history (Mason-Gamer *et al.* 1995).

The 273 chloroplast genomes analyzed revealed 33 variable restriction sites, and 13 distinct haplotypes. The estimated sequence divergence between the most divergent haplotypes was nearly 0.7%. Using methods described in Holsinger and Mason-Gamer (1996), it is possible to construct a tree that represents presumed population relationships based on the average coalescence time of haplotypes within each population sample (Fig. 2.2). The major divergence identified at the base of the tree in Figure 2.2 corresponds with neither the geographic distinction between east and west nor the partitioning of populations into morphological varieties. Rather, it corresponds with a previously unsuspected

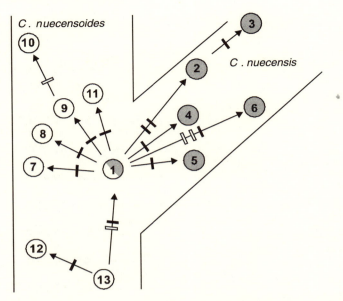

Figure 2.1. Distribution of haplotypes among *Coreopsis nuecensis* and *C. nuecensoides.* Open circles represent haplotypes found in *C. nuecensoides.* Striped circles represent haplotypes found in *C. nuecensis.* Closed hatch marks on arrows represent loss of a restriction site. Open hatch marks represent gain of a restriction site. (Redrawn from Mason-Gamer *et al.* 1999.)

divergence between two broad classes of chloroplast genomes, referred to as A and B genomes.

Within each genome class, however, the geographic distinction is primary. Since chloroplast DNA is usually inherited maternally (but see Mason *et al.* 1994 for documentation of a rare exception in this species), these results mean that A-genome populations in the eastern part of the distribution, northwestern Georgia, actually share a more recent maternal ancestor with A-genome populations in Arkansas than they do with B-genome populations that also occur in Georgia. This is particularly striking because both one B genome population in Georgia and one B-genome population in Arkansas are subsets of B genomes found in single populations that contained both A and B genomes. Within these populatons, therefore, two very different maternal lineages are represented. This study makes it clear that cpDNA markers have substantial potential for addressing systematic questions at the population or species level in plants. They have the potential both to identify evolutionary units

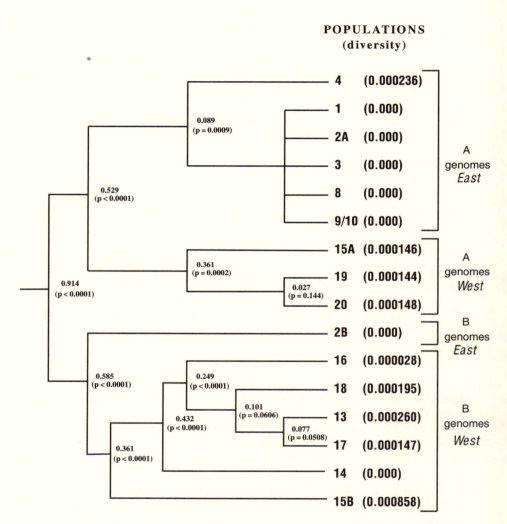

Figure 2.2. Hierarchical analysis of nucleotide diversity in *Coreopsis grandiflora*. The number at each node is the evolutionary distance between its two daughter nodes, and the *P*-value reported is the probability of obtaining a distance greater than the observed distance under the null hypothesis of no genetic differentiation (based on 10,000 random resamplings). (Redrawn from Holsinger and Mason-Gamer 1996.)

and to identify evolutionary divergences that might not have previously been expected.

Evolutionary Patterns in North American *Crepis*

Although they have been little used for this purpose, cpDNA markers may also provide some evidence of population connections in a weak demographic sense. The genus *Crepis* was the subject of a classic series of studies by E. B. Babcock and G. L. Stebbins in the 1930s and 1940s (summarized in Babcock and Stebbins 1938). As a result of studies in this genus, ideas about the nature of polyploid and apomictic complexes were first fully developed. In the first detailed study of this group since the 1940s, Whitton (1994) used cpDNA restriction-site variation to determine the relative importance of independent evolution at the polyploid level and multiple origins of polyploid apomicts to determine the patterns of genetic diversity currently found within the complex.

Analysis of samples from forty-five populations throughout western North America revealed forty-five variable restriction sites and sixteen distinct haplotypes. The pattern revealed is striking. If genetic diversity among apomicts reflected independent evolution at the polyploid level, diagnosable morphological differences used to recognize species within the apomictic complex would correspond with monophyletic haplotype lineages. They do not (Fig. 2.3). Instead, it appears that much of the genetic diversity found in the apomictic complex results from multiple, independent origins of apomictic lines, not from diversification at the polyploid level.

A second important implication follows from the geographic distribution of morphotypes and haplotypes. Five populations contain both more than one morphotype and more than one haplotype lineage; for example, population 153 contains both morphotypes corresponding to *Crepis acuminata* and *C. modocensis* and haplotypes 4 and 6. Moreover, many haplotypes occur in more than one population. It seems likely that populations containing more than a single morphotype or more than a single haplotype represent populations in which seed dispersal has led to the establishment of different maternal lines in a single populations, as with the divergent maternal lineages discovered in two populations of *Coreopsis grandiflora*. Populations 153 and 159, for example, have haplotypes within them with very different maternal histories. Because these plants are primarily apomictic, it appears that their entire history, not just their maternal ancestry, is relatively distinct. Thus, seed dispersal between existing populations appears to have contributed significantly to the genetic diversity within apomictic populations of *Crepis* in western North America. Little of the detectable genetic diversity within

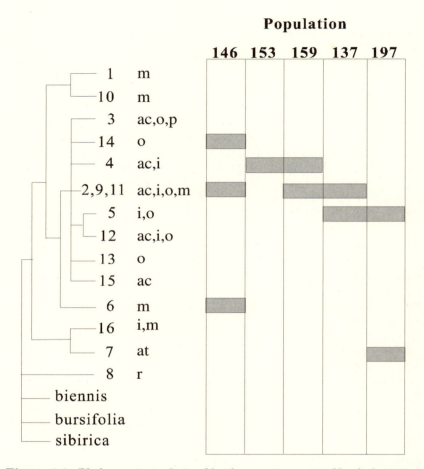

Figure 2.3. Phylogenetic analysis of haplotypes in western North American *Crepis*. Numbers designate different haplotypes. Letters indicate the morphotypes having the corresponding haplotype. Gray bars indicate the haplotypes present in the subset of five populations illustrated here. (Adapted from Whitton 1994.)

populations is a result of genetic differences within a single morphotype within populations. The genetic diversity within populations is mostly the result of mixing evolutionarily independent populations.

Limits of Molecular Markers

The studies of *Coreopsis* and *Crepis* illustrate some of the ways in which cpDNA markers may be useful at the population level for answering evolutionary questions. They suggest how cpDNA markers could be used either to provide insight into the degree of demographic connection between isolated endangered plant populations or to identify evolutionarily independent plant populations that are worthy of conservation. Having illustrated their potential utility, however, it is also important to understand some of their limits.

The limits of molecular markers derive from several different sources (Lynch 1996). First, the evolutionary dynamics of molecular markers and adaptive traits are likely to be quite different simply because the rate at which new variants are introduced is also quite different. Nucleotide sequence and restriction-site variants are introduced at a rate of 10^{-8} to 10^{-5} per generation. New polygenic variants, on the other hand, arise at a rate of 10^{-4} to 10^{-3} per generation. Thus, the rate at which new variants arise is two to three, possibly even five, orders of magnitude greater for polygenic traits than it is for those traits that we measure for molecular variation. Adaptive responses to future environmental conditions will often have a polygenic basis. As a result, patterns of genetic diversity revealed through an analysis of molecular variants may be quite different from the patterns that will determine the future success and long-term viability of populations. In particular, additive genetic variance in polygenic traits will rebound much more rapidly following a population bottleneck than will diversity of molecular markers.

Second, heterozygosity for molecular markers decreases linearly with increases in inbreeding. In the presence of significant nonadditive allelic or epistatic interactions, however, additive genetic variance may increase as a result of inbreeding, especially if the inbreeding results from a population bottleneck. Population bottlenecks reduce heterozygosity for molecular markers and lead to loss of rare alleles, but they may, under a variety of circumstances, increase the ability of populations to respond to future environmental change, at least in the short term. Again, the evolutionary dynamics of molecular markers from which we hope to gain insight into population processes may be quite different from the dynamics of genetic variation that determines their long-term persistence.

Third, in populations that are close to panmixis, alleles at different loci are expected to show similar patterns of variation only if they are subject to similar selection pressures. Coalescent theory, which has been extensively developed in the past fifteen years, has conclusively shown that the genealogical relationship among a set of alleles depends strongly on the nature and magnitude of selective forces acting on them (see Hudson

1990 for an accessible review of the mathematical theory). One consequence of this general result is that genealogical relationships among a set of alleles at unlinked loci will become congruent only after populations have stopped exchanging genes for a long period of time (Avise and Ball 1990; Avise 1994, 1995; Holsinger and Vitt 1997). In fact, allelic genealogies may remain discordant for a very long time if alleles are subject to some form of balancing selection. Frequency-dependent selection for rare self-incompatibility alleles, for example, has resulted in extensive trans-specific polymorphisms. The $S2$ allele of a close relative of cultivated potato (*Solanum chacoense*) is more closely related to the *Sa* allele of a tobacco relative (*Nicotiana alata*) than it is to the $S11$ allele, which is also found in *S. chacoense* (Richman *et al.* 1996).

Molecular Markers and the Dusky Seaside Sparrow

The dusky seaside sparrow is an extinct subspecies of seaside sparrow that once was the subject of intense efforts by the United States Fish & Wildlife Service to prevent its extinction. The population was known only from a small part of Brevard County, Florida, and, as suggested by its name, its appearance was quite different from all other recognized subspecies of seaside sparrow. After the last remaining male died a few years ago, Avise and Nelson (1989) extracted DNA from that individual and from representatives from other populations and subspecies of the seaside sparrow throughout its geographic range on the Atlantic and Gulf coasts of the United States. They found a variety of different mitochondrial DNA (mtDNA) haplotypes, and they documented a major genetic divergence between Atlantic and Gulf coast populations. Significantly, the one individual of dusky seaside sparrow included in their sample was only marginally distinct from other populations along the Atlantic Coast. They suggested that conservation efforts might have been better directed at protecting multiple populations of seaside sparrow along both the Atlantic and Gulf coasts than at trying to prevent extinction of a single population that was, at best, marginally distinct.

Mitochondrial DNA is, however, maternally inherited. As a result, the conservation implications of this study depend on whether we believe that the maternal phylogeny of the last surviving male dusky seaside sparrow, which Avise and Nelson (1989) studied, is concordant with the population phylogeny. If this individual had *any* maternal ancestor belonging to one of the nearby Atlantic Coast populations, his mtDNA would reflect that ancestry even if the population of which he was the last representative was evolutionarily very distinct. Even if the hybridization event involving this maternal ancestor occurred tens or hundreds of generations ago, his maternal phylogeny would not reflect that of the population to which he once belonged.

To see why, consider another part of the results from *Coreopsis grandiflora* (Mason-Gamer *et al.* 1995; Fig. 2.4). The *A* and *B* genomes so far identified in *Coreopsis* differ from one another by a minimum of ten restriction sites. The differences within genomes are far smaller. *Coreopsis nuecensis* and *C. nuecensoides*, for example, carry only the *B* genome, and the most common haplotype in each differs from the common *B* genome in *C. grandiflora* at only one of the scored restriction sites. Limited population samples from other members of section *Coreopsis* indicate that each seems to carry either the *A* genome or the *B* genome, although *C. lanceolata* may share the *A/B* polymorphism with *C. grandiflora*. Even species outside section *Coreopsis* do not seem to carry a genome that is clearly ancestral to both *A* genomes and *B* genomes. Members of some sections appear to carry the *A* genome, while members of other sections carry the *B* genome. The extensive trans-specific polymorphism for plastid genomes suggested by these results is consistent with an ancient hybridization event, after which both genomes have continued to exist in at least one species for a very long time.

Suppose hybridization occured only once between a male dusky seaside sparrow and a female from one of the other Atlantic Coast populations, and suppose that this female was the direct maternal ancestor of the sampled individual. Then the mtDNA sample obtained from this individual would be expected to differ only slightly from that of other Atlantic Coast subspecies. Its nuclear genome, however, might be very distinct. Its nuclear genome could easily have been assimilated to the resident dusky seaside sparrow's genetic background through repeated backcrossing. Had mtDNA haplotypes of many different dusky seaside sparrow individuals shown the same pattern of relationships, however, a close mtDNA relationship among sampled populations represented could also be interpreted as strong evidence for a close relationship at the population level.

2.4 CONCLUSIONS: THE FUTURE OF PLANT CONSERVATION GENETICS

Genetics has two important contributions to make to future plant conservation efforts. First, it can contribute techniques that help to identify evolutionarily independent populations worthy of conservation. Second, it can contribute techniques that provide insight into demographic processes that are otherwise difficult or impossible to study.

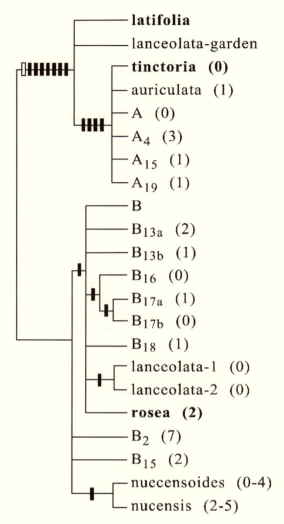

Figure 2.4. Phylogenetic analysis of haplotypes from *Coreopsis grandiflora* and other members of the genus. The tree is midpoint rooted because no appropriate outgroup has yet been identified. Taxa in boldface belong to other sections of the genus *Coreopsis*. Filled bars correspond to unique changes, the open bar to a site showing a reversal. Numbers in parentheses indicate the minimum number of restriction-site changes along each terminal branch.

Identifying evolutionarily independent populations will require not only the development of new molecular markers that allow finer evolutionary distinctions to be made, but will also require the design of appropriate sampling strategies to detect that variation. The need for new molecular markers should be apparent. The need for new sampling strategies may become so after a bit of reflection. Extensive population surveys may reveal relatively little differentiation among populations within species, as in *Crepis*, or they may unexpectedly reveal a large genetic divergence between genomes in populations of the same species, as they did in *Coreopsis grandiflora*. Until the surveys are done, however, we have no way of guessing whether significant evolutionary structure might be revealed. We desperately need new sampling techniques that allow us rapidly to identify those populations that might be evolutionarily distinct.

Providing insight into demographic processes is a task of a different sort, but the evolutionary analysis of *Crepis* shows that analysis of cpDNA variation may provide insight into patterns of seed migration among populations. Another promising example is provided by the development of a rapid molecular assay for self-incompatibility genotypes in members of the potato family (Richman *et al.* 1996). To estimate the number of self-incompatibility alleles segregating in a population, all pairwise crosses between a large subset of individuals in the population must be performed. For a sample of twenty individuals, a total of 380 separate crosses, each with several replicates, is necessary. The reverse-transcriptase polymerase-chain-reaction (RT-PCR) technique Richman *et al.* (1996) developed is quite straightforward, at least in principle:

1. Extract messenger RNA from styles, because the messenger RNA from the self-incompatibility locus is greatly enriched in styles.
2. Reverse transcribe the extracted RNA into cDNA with reverse-transcriptase.
3. Amplify the cDNA through the polymerase chain reaction (PCR) with primers specific for the S locus.

The two PCR products derived from this RT-PCR protocol satisfy all properties expected of a molecular product of the S locus. Use of this technique allows individuals to be genotyped quite rapidly. If allele specific primers can be developed for these stylar RNAs, even more rapid genotyping may be possible, far more rapid than with the complicated series of reciprocal crosses that is otherwise necessary to identify all the self-incompatibility alleles segregating within a population.

Caughley and the Two Paradigms of Conservation Biology

Over the last fifteen years two paradigms have coexisted uneasily in conservation biology. The small-population paradigm (Caughley 1994), which dominated conservation biology in the 1980s, is concerned primarily with managing threats to populations that are already small. Because those populations are small, many of the problems associated with management are problems associated with variation in population dynamics or genetic composition. Because that variation is related primarily to population size rather than to life history, it has been possible to develop a broadly applicable, widely useful mathematical theory for the properties of those populations. That theory has been and continues to be developed.

While the small-population paradigm developed from the increasing academic interest in conservation problems, practical people involved with endangered species management, partly because they were not located in academic departments and partly because of their differences in training, have focused primarily on mitigating or reversing threats to continued population persistence. They practice within what Caughley (1994) referred to as the declining-population paradigm. The emphasis among these practitioners has been not on stochastic threats to persistence of small populations, but on the deterministic processes that have caused populations to be small in the first place. Unfortunately, the specific causes of decline seem to differ from one species to another so much that it may be that no general theory is possible within the declining-population paradigm. It certainly seems to be true that no general theory has yet emerged from within that paradigm.

The two most important lessons learned from the small-population paradigm, however, are that managing small populations requires an enormous effort and that long-term persistence of populations is best ensured by helping them rapidly to become as large as possible (Holsinger and Vitt 1997). Enormous investments of time, money, and other resources are required for long-term maintenance of small populations. As a result, the best chance population managers have to ensure long-term persistence of the populations under their care is to focus on reversing the deterministic threats those populations face and to do whatever is necessary to increase the population size rapidly so that the stochastic threats are minimized.

As a result of this focus on deterministic threats, the role of genetics in endangered species management is relatively limited. There are only two cases in which the genetic structure of a plant population is itself a direct threat to its continued persistence:

1. When self-incompatibility alleles have been lost, the reproductive capacity of the population may be directly threatened.

2. If a rare species occurs in close contact with a reproductively compatible relative, hybridization may threaten extinction through genetic assimilation.

Direct management of the genetic structure may be necessary only rarely, but this is not to suggest that conservation managers can ignore genetics. Rather, it is to suggest that many of the recommendations for management may be the same whether they are derived from principles of genetics, systematics, evolutionary biology, biogeography, or ecology. It makes sense on the basis of both genetic and ecological principles, for example, to design preserve systems that maintain multiple, independent populations of a endangered species throughout most of their geographic range. The reasons a geneticist gives for that advice may be different from those an ecologist would provide, but the practical impact is the same.

The distinctive contribution conservation genetics can make to endangered plant conservation is more limited than many of us thought ten or fifteen years ago. What genetics can do is to provide important tools for the kit of conservation biologists. In providing these tools, we cannot forget that the huge number of plant species facing extinction precludes providing individual attention to more than a tiny fraction of them. Furthermore, the threats to the remaining native ecosystems increase every day and, as Lande (Chapter 1 in this volume) emphasizes, the primary source of these threats is to be found in habitat loss, habitat degradation, and overexploitation of species, not in genetic processes. If generations to come are to find a world as rich and full of diversity as our own, we must be as aware of the economic, sociological, and political contexts in which conservation principles are applied as we are of how to derive those principles from genetics and ecology.

ACKNOWLEDGMENTS

Don Stratton and two anonymous reviewers read an earlier version of this chapter and provided suggestions that improved it substantially. Portions of the work described here benefited from support by the National Science Foundation (BSR-9105167 to R. J. Mason-Gamer; BSR-9212989 to J. Whitton; DEB-9509006 to K. E. Holsinger) and the University of Connecticut Research Foundation.

References

Aitken, S. N. and Libby, W. L. (1994). Evolution of the pygmy-forest edaphic subspecies of *Pinus contorta* across an ecological staircase. *Evolution*, *48*, 1009–1019.

Avise, J. C. (1994). *Molecular Markers, Natural History, and Evolution*. New York: Chapman and Hall.

Avise, J. C. (1995). Mitochondrial DNA polymorphism and a connection between genetics and demography of relevance in conservation. *Conserv. Biol.*, *9*, 686–690.

Avise, J. C. and Ball, J.R.M. (1990). Principles of genealogical concordance in species concepts and biological taxonomy. *Oxford Surv. Evol. Biol.*, *7*, 45–67.

Avise, J. C. and Nelson, W. S. (1989). Molecular genetic relationships of the extinct dusky seaside sparrow. *Science*, *243*, 646–648.

Babcock, E. B. and Stebbins, J.G.L. (1938). The American species of *Crepis*. No. 504. Washington, D.C.: Carnegie Institute.

Banks, J. A. and Birky, J.C.W. (1985). Chloroplast DNA diversity is low in a wild plant, *Lupinus texensis*. *Proc. Natl. Acad. Sci. USA*, *82*, 6950–6954.

Bradshaw, A. D. (1984). Ecological significance of genetic variation between populations. In Dirzo, R. and Sarukhan, J. (eds.), *Perspectives on Plant Population Ecology*, 213–228, Sunderland, Mass.: Sinauer.

Bürger, R. and Lynch, M. (1995). Evolution and extinction in a changing environment: A quantitative-genetic analysis. *Evolution*, *49*, 151–163.

Cade, T. J. (1983). Hybridization and gene exchange among birds in relation to conservation. In Schoenwald-Cox, C. M., Chambers, S. M., MacBryde, B., and Thomas, W. L. (eds.), *Genetics and Conservation: A Reference for Managing Wild Animal and Plant Populations*, 288–310, Menlo Park, CA: Benjamin-Cummings.

Caughley, G. (1994). Directions in conservation biology. *J. of Animal Ecol.*, *63*, 215–244.

Cosner, M. B. and Crawford, D. J. (1990). Allozyme variation in *Coreopsis* sect. *Coreopsis* (Asteraceae). *Syst. Bot.*, *15*, 256–265.

Couland, J. and McNeilly, T. (1992). Zinc tolerance in populations of *Deschampsia caespitosa* (Gramineae) beneath electricity pylons. *Plant Syst. Evol.*, *179*, 175–185.

Crawford, D. J., Palmer, J. D., and Kobayashi, M. (1990). Chloroplast DNA restriction site variation and the phylogeny of *Coreopsis* section *Coreopsis*. *Am. J. Bot.*, *77*, 552–558.

Crawford, D. J. and Smith, E. B. (1982). Allozyme variation in *Coreopsis nuecensoides* and *C. nuecensis* (Compositae), a progenitor-derivative species pair. *Evolution*, *36*, 379–386.

Crawford, D. J. and Smith, E. B. (1984). Allozyme divergence and interspecific variation in *Coreopsis grandiflora* (Compositae). *Syst. Bot.*, *9*, 219–225.

Crow, J. F. and Kimura, M. (1970). *An Introduction to Population Genetics Theory*. Burgess Publishing Company, Minneapolis, MN.

DeMauro, M. M. (1993). Relationship of breeding system to rarity in the lakeside daisy (*Hymenoxys acaulis* var. *glabra*). *Conserv. Biol.*, *7*, 542–550.

Doyle, J. J. and Doyle, J. L. (1987). A rapid DNA isolation procedure for small quantities of fresh leaf tissue. *Phytochem. Bull.*, *19*, 11–15.

Ellstrand, N. C. (1992). Gene flow by pollen: Implications for plant conservation genetics. *Oikos*, *63*, 77–86.

Ellstrand, N. C. and Elam, D. R. (1993). Population genetic consequences of small population size: implications for plant conservation. *Ann. Rev. Ecol. Syst.*, *24*, 217–242.

Ewens, W. J. (1979). *Mathematical Population Genetics*. Berlin: Springer-Verlag.

Ewens, W. J., Brockwell, P. J., Gani, J. M., and Resnick, S. I. (1987). Minimum viable population size in the presence of catastrophes. In Soulé, M. E. (ed.), *Viable Populations for Conservation*, Cambridge: Cambridge University Press.

Frank, S. A. and Slatkin, M. (1992). Fisher's fundamental theorem of natural selection. *Trends Ecol. Evol.*, 7, 92–95.

Holsinger, K. E. (1995a). Conservation programs for endangered plant species. In Nierenberg, W. A. (ed.), *Encyclopedia of Environmental Biology*, Vol. 1, 385–400, San Diego: Academic Press.

Holsinger, K. E. (1995b). Population biology for policy makers: Promises and paradoxes. *BioScience, 45(Supplement)*, S10–S20.

Holsinger, K. E. and Gottlieb, L. D. (1991). Conservation of rare and endangered plants: Principle and prospects. In Falk, D. A. and Holsinger, K. E. (eds.), *Genetics and Conservation of Rare Plants*, 195–208. New York: Oxford University Press.

Holsinger, K. E. and Mason-Gamer, R. J. (1996). Hierarchical analysis of nucleotide diversity in geographically structured populations. *Genetics, 142*, 629–639.

Holsinger, K. E. and Vitt, P. (1997). The future of conservation biology: What's a geneticist to do? In Pickett, S. T. A., Ostfeld, R. S., Shachak, M., and Likens, G. E. (eds.), *The Ecological Basis of Conservation: Heterogeneity, Ecosystems, and Biodiversity*, 202–216, New York: Chapman and Hall.

Hudson, R. R. (1990). Gene genealogies and the coalescent process. *Oxford Surv. Evol. Biol.*, 7, 1–44.

Jansen, R. K. and Palmer, J. D. (1987). Chloroplast DNA from lettuce and *Barnadesia* (Asteraceae): structure, gene localization, and characterization of a large inversion. *Curr. Genet.*, 11, 553–564.

Kim, K.-J., Jansen, R. K., and Turner, B. L. (1992). Phylogenetic and evolutionary implications of interspecific chloroplast DNA variation in dwarf dandelions (*Krigia*; Asteraceae). *Syst. Bot.*, 17, 449–469.

Kirkpatrick, M. and Barton, N. (1997). Evolution of a species' range. *Am. Natur.*, 150, 1–23.

Lande, R. (1988). Genetics and demography in biological conservation. *Science, 241*, 1455–1460.

Lande, R. (1994). Risk of population extinction from fixation of new deleterious mutants. *Evolution, 48*, 1460–1469.

Lande, R. (1995). Mutation and conservation. *Conserv. Biol.*, 9, 782–791.

Lande, R. and Shannon, S. (1996). The role of genetic variation in adaptation and population persistence in a changing environment. *Evolution, 50*, 434–437.

Lynch, M. (1996). A quantitative-genetic perspective on conservation issues. In Avise, J. C. and Hamrick, J. L. (eds.), *Conservation Genetics: Case Histories from Nature*, 471–501, New York: Chapman and Hall.

Lynch, M., Conery, J., and Bürger, R. (1995). Mutation accumulation and the extinction of small populations. *Am. Natur.*, 146, 489–518.

Lynch, M. and Lande, R. (1993). Extinction and evolution in response to environmental change. In Kareiva, P. M., Kingsolver, J. G., and Huey, R. B. (eds.), *Biotic Interactions and Global Change*, 234–250, Sunderland, Mass.: Sinauer.

Mason, R. J., Holsinger, K. E., and Jansen, R. K. (1994). Biparental inheritance of the chloroplast genome in *Coreopsis* (Asteraceae). *J. Hered.*, 84, 171–173.

Mason-Gamer, R. J., Holsinger, K. E., and Jansen, R. K. (1995). Chloroplast DNA haplotype variation within and among populations of *Coreopsis grandiflora* (Asteraceae). *Mol. Biol. Evol.*, 12, 371–381.

Mason-Gamer, R. J., Holsinger, K. E., and Jansen, R. K. (1999). Chloroplast DNA variation in *Coreopsis nuecensoides* and *C. nuecensis* (Asteraceae), a presumed progenitor-derivative species pair. *Plant Syst. Evol.*, (in press).

Muller, H. J. (1964). The relation of recombination to mutational advance. *Mutation Res.*, *1*, 2–9.

Raven, P. H. (1987). The scope of the plant conservation problem world-wide. In Bramwell, D., Hamann, O., Heywood, V., and Synge, H. (eds.), *Botanic Gardens and the World Conservation Strategy*, 19–29, London: Academic Press.

Richman, A. D., Uyenoyama, M. K., and Kohn, J. R. (1996). Allelic diversity and gene genealogy at the self-incompatibility locus in the solanaceae. *Science*, *273*, 1212–1216.

Rieseberg, L. H. (1991). Hyrbridization in rare plants: Insights from case studies in *Cercocarpus* and *Helianthus*. In Falk, D. A. and Holsinger, K. E. (eds.), *Genetics and Conservation of Rare Plants*, 171–181, New York: Oxford University Press.

Rieseberg, L. H. and Gerber, D. (1995). Hybridization in the catalina island mountain mahogany (*Cercocarpus traskiae*): RAPD evidence. *Conserv. Biol.*, *9*, 199–203.

Schmitt, J. and Gamble, S. E. (1990). The effect of distance from the parental site on offspring performance and inbreeding depression in *Impatiens capensis*: A test of the local adaptation hypothesis. *Evolution*, *44*, 2022–2030.

Smith, E. B. (1974). *Coreopsis nuecensis* and a related new species from southern texas. *Brittonia*, *26*, 161–171.

Soltis, D. E., Soltis, P. S., and Milligan, B. G. (1992). Intraspecific chloroplast DNA variation: Systematic and phylogenetic implications. In Soltis, P. S., Soltis, D. E., and Doyle, J. J. (eds.), *Molecular Systematics of Plants*, 117–150, New York: Chapman and Hall.

Stanton, M. L., Galen, C., and Shore, J. (1997). Population structure along a steep environmental gradient: consequences of flowering time and habitat variation in the snow buttercup, *Ranunculus adoneus*. *Evolution*, *51*, 79–94.

Stratton, D. A. (1994). Genotype-by-environment interactions for fitness of *Erigeron annuus* show fine-scale selective heterogeneity. *Evolution*, *48*, 1607–1618.

Walter, K. S. and Gillett, H. J. (1998). *1997 IUCN Red List of Threatened Plants*. Cambridge: International Union for the Conservation of Nature and Natural Resources — The World Conservation Union.

Whitton, J. (1994). Systematic and evolutionary investigation of the North American *Crepis* agamic complex. Ph.D. dissertation, University of Connecticut, Storrs.

3

Genetic Theory and Evidence Supporting Current Practices in Captive Breeding for Conservation

KATHRYN M. RODRÍGUEZ-CLARK

SUMMARY. This chapter begins with a discussion of the principles that underlie the basic strategy for retaining neutral genetic variation in captive-bred populations. I then present the theory behind recent attempts to improve upon this general strategy by using pedigree information, giving measures of genetic importance such as "mean kinship." This, in principle, allows optimal retention of genetic diversity by giving high priority to animals that have the lowest degree of relatedness to all other members of the living population. Next, I review several experiments that have tested components of the general strategy, and a few that have focused specifically on mean kinship. These experimental tests have focused, disappointingly, on proximate outcomes (slowing loss of neutral variation and/or preventing inbreeding depression), leaving the ultimate goal of retention of adaptive potential unexamined.

I then shift focus to more fundamental research, reviewing the theory, assumptions, and evidence behind the classical view that loss in population heterozygosity and thus neutral variation should be proportional to loss in adaptive variation. I first examine the merits of correlational evidence from wild populations, and then review experimental studies in laboratory populations, which are largely studies of *Drosophila*. While sharply divided, these studies generally support both the classical view and thus the mean kinship approach, though more direct evidence is needed. They ultimately point to important aspects of experimental design required of future studies, which must expand beyond the realm of *Drosophila*.

INTRODUCTION

In the past fifteen years, the focus of zoological parks worldwide has moved dramatically towards conservation as a central goal (Ballou and

Foose 1996). With this shift has come a rapid intensification of attention paid to breeding practices. The number of species with pedigrees managed in accordance with principles in population genetics and demography has jumped from a handful in 1965 to over 250 in 1996 (Ballou and Foose 1996; Shoemaker and Flesness 1996). While conservation is often achieved most effectively through protection of habitat and wild populations, for many species captive breeding is the only option at present (Magin *et al.* 1994; Snyder *et al.* 1996).

Many parameters may affect the success of a captive breeding effort; these include the behavioral, social, physiological, and even physical characters of the species in question, as well as the administrative, economic, legal, and political characteristics of institutions involved (Beck *et al.* 1994; Caughley and Gunn 1996; Kleiman *et al.* 1996 and chapters therein). The focus of the following review is thus neccesarily quite narrow, restricted to the genetic aspects of captive breeding, and then only to the theory as applied to species that can be individually managed. When individual management is not possible, other strategies may be more applicable (Dobson *et al.* 1992; Lacy *et al.* 1995; Princée 1995), although in theory the strategies outlined below may be extended to group management (Lacy, pers. comm.). Similarly, the significant practical problems often faced in actually implementing breeding strategies must be left to other reviewers (i.e., Woodruff 1989).

Basic Problems, Solutions, and Research Needs

From a genetic perspective, then, breeding practices are particularly important for their impact on the genetic problems faced by "closed populations" — species that must be self-sustaining in captivity, with no future inputs from wild stock. The conception of these problems has changed little since they were first enunciated by Franklin (1980) and Soulé (1980). The first type of problem involves changes in the *mean* value of characters — either selected shifts, which occur during adaptation to captivity, or undesired decreases, which are known as inbreeding depression. The second type of problem involves changes in the *variability* of characters, particularly loss of variability, which may impair species' ability to adapt to changing environments. The overall goal of captive breeding in closed populations is to prevent these problems by "minimiz[ing] change in the genetic constitution of the population while in captivity so that if and when the opportunity arises for animals to be reintroduced into the wild, they will represent, as closely as possible, the genetic characteristics of the original founders used to establish the population" (Ballou and Foose 1996, p.264). Thus, the goal is not maximal variability *per se*, but maximum retention of *existing* variability.

This goal of maximal retention implies that the rate of future evolution in a population will be variation-limited. Lande (1988) has rightly pointed out one weakness of this viewpoint: in natural populations, evolution (at least in morphological characters) appears to be limited largely by the strength of natural selection itself (e.g., Simpson 1953). However, this reasoning ignores the fact that current rates of environmental change are causing extinctions several orders of magnitude faster than any mass extinction in the past (May *et al.* 1995), potentially shifting the balance of rate-limiting factors. Although the optimum level of genetic variation depends on the extent of environmental fluctuations, Lande and Shannon (1996) conclude that the environmental shifts likely to occur under human disturbance will make maintenance of normal levels of genetic variation increasingly important for long-term population persistence.

Various strategies have been developed in recent years to meet the goal of maximal retention of variation, based largely on the theory (developed below) that slowing the loss of neutral molecular heterozygosity is the best means of both preventing inbreeding depression and preserving genetic variability. Empirical testing has not kept pace with changing practices, however. Only one study has attempted a partial experimental evaluation of the breeding strategy recommended currently for all intensively managed Association of American Zoos and Aquariums–pedigreed populations (see below; Montgomery *et al.* 1997).

Such empirical investigation is important for several reasons. In the first place, the fate of genetic variation in small or fluctuating populations has been a central concern in evolutionary biology (e.g., Wright 1931), and is fundamental to our understanding of mechanisms of evolution in nature. The interactions of genes and the genetic processes of random drift and selection in small populations have been postulated to play a major role in speciation (e.g., Mayr 1963; Carson 1975; Templeton 1989; Hollocher 1996), although exact mechanisms remain controversial (Barton and Charlesworth 1984; Tauber and Tauber 1989; Provine 1989; Rice 1993). Furthermore, while breeding programs have been designed to maintain neutral molecular genetic heterozygosity, the features of organisms that are actually important to their survival and persistence are by definition not neutral. These adaptive traits are often influenced by tens or hundreds of factors and may vary continuously; they are thus referred to as "quantitative traits" (Lynch 1996). The relationship between factors influencing neutral heterozygosity and those affecting quantitative variation remains unclear (see below).

In addition, empirical research provides practical information to aid zoo management. Logistics, legal constraints on the movement of animals, embryos, and gametes, and interinstitutional agreement may be among the most difficult aspects of captive breeding; if managing a population "perfectly" takes considerably more effort than more laissez-faire approaches, the exact benefits of perfect management need to be known; are they enough to justify the effort invested? While some individual components of an optimal breeding strategy have been tested and found robust for proximate goals (see below), they have not been brought together to ensure that proximate successes combine as expected to achieve ultimate goals. Finally, the vast majority of these studies have been conducted with a single species, *Drosophila melanogaster* (Ralls and Meadows 1993, and references below). Results outside this species will be desirable to demonstrate that breeding methods are as widely applicable as they are presumed to be.

3.1 PRESERVING VARIATION

Least Favored Method: "Active Selection"

Initial breeding practices in zoos were adopted directly from agriculture. Animals were either left to themselves, and required continual supplement from the wild to maintain populations, or they were selectively bred, using only the "best" animals, thus encouraging domestication. This "active selection" strategy has largely been abandoned in zoos. However, it has modern counterparts, which may be appropriate where conservation is not a primary goal, or when removing extreme mutants among siblings is necessary to reduce genetic load (Arnold 1995; Ballou and Foose 1996).

Intentional selection has also been advocated in some instances for conservation purposes. For example, selection for rare Major Histocompatibility Complex (MHC) alleles in mammals has been suggested (Hughes 1991). This has been strongly criticized for a number of reasons, however. First, proving a direct relationship between MHC variability and the immune response with which it is presumed to be related is, in fact, difficult (Hedrick 1996). Second, conducting selection at MHC loci assumes that variability at other loci (which is often lost in selective regimes) is irrelevant—a view that is highly questionable (Gilpin and Willis 1991; Vrijenhoek and Leberg 1991). Finally, the uneven distribution of rare alleles in populations often makes it impossible to increase the frequency of several rare alleles simultaneously, so that rare alleles may be lost by the "MHC manager" anyway (Hedrick and Miller 1994). The potential loss of variation under any selective regime is a general phenomenon; for example, the original global policy on genetic management of Przewalski's horses (*Equus przewalskii*) was to not mate individuals descended from a known Mongolian domestic mare, which was unfortunately bred into the pedigree in an early generation. Subsequent analysis has shown, however, that due to the structure of the species' pedigree, this policy would have resulted in the loss of about 25% of the original founder genes (Thompson 1995).

Preserving Variation: Basic Principles

Given that selective methods are underdeveloped at best, the most conservative strategy would be to immortalize the founding individuals, without presuming to know which will be most important to the species' future survival (Ballou 1984). Maximizing generation length approaches this strategy but obviously cannot reach it: finite breeding populations inevitably lose variability through time. When genetic variability is defined as heterozygosity (Nei 1978), as it often is (Allendorf 1986), the way in which these losses occur is well understood. It should be noted that the term heterozygosity is used in the genetic literature (and in keeping with that, below) to indicate two slightly different concepts, usually clear from the context; the first is simply the observed proportion of individuals that have two different alleles at a given locus, or that proportion averaged across a number of loci. The second meaning, more accurately described as "gene diversity" (Lacy 1995), is the expected proportion of heterozygotes at a locus in a randomly mating population. This is given by $1 - \sum_i p_i^2$, where p is the frequency of an allele at that locus, and the sum is over all alleles from 1 to i. In a randomly mating, ideal population (as in Hardy 1908; Weinberg 1963), expected and observed heterozygosities are equivalent.

Six processes govern changes in heterozygosity: mutation, migration, drift, subdivision, inbreeding, and selection (Wright 1977a; recently reviewed by Amos and Harwood 1998). Mutation acts primarily to increase heterozygosity (Kimura and Crow 1964; Lande 1976), while migration can cause either increases or decreases, depending on the similarities between donor and recipient populations (Falconer and MacKay 1996). The combination of random segregation of genes into gametes and unequal reproduction among individuals due to environmental factors produces random changes in allele frequencies, otherwise known as drift. As populations get smaller, this sampling error increases and leads to increasing losses of rare alleles and thus a predictable decline in heterozygosity (Wright 1931; Kimura and Crow 1964). Subdivision, in creating smaller groups within which genes are exchanged, also increases the rate of drift within each subpopulation. Similarly, heterozygosity is lost with systematic inbreeding simply because individuals mate with relatives and thus are more likely to contribute the same allele at a given locus than would occur by chance.

Finally, the effect of selection varies greatly depending on the particulars of its operation. Sustained directional selection can reduce heterozygosity (Whitlock and Fowler 1999), while the reverse is true in situations involving heterozygote superiority (overdominance), frequency-dependent selection (Falconer and MacKay 1996), spatially fluctuating selection (Levene 1953), temporally fluctuating selection (Haldane and Jayakar 1963), or antagonistic pleiotropy (Rose 1982).

In order to study the simultaneous action of these forces in a given population, a common "currency" has been derived, the *inbreeding effective population size*. This population size (N_e), is that number of ideal, randomly mating individuals that would lose neutral heterozygosity at the same rate as the population under question (Wright 1931; Crow and Kimura 1970). The definitions of "random" and "ideal," in this case, are particular ones. Offspring are "assigned" to randomly-chosen pairs of parents (who are assumed to be hermaphroditic and self-compatible). The resulting distribution of family sizes for each pair of parents is a Poisson distribution, with a mean of two offspring per family and a variance of two. Thus, there will be a predictably small number of families with zero offspring, and regular loss of heterozygosity through time. Inbreeding effective size is closely related to another measure, the variance effective size. These are identical when mating is random and population size is constant; otherwise the former is more dependent on parental generation size, while the latter is more dependent on the offspring generation size (Kimura and Crow 1963b; Crow and Kimura 1970, p.357). In either case, loss of variation is slower when N_e is larger. As Chapters 1 and 2 point out, in each generation the loss due to drift is simply $1/(2N_e)$ (Kimura and Crow 1964).

Effective population size, therefore, depends not only on the actual number of individuals present, but also on their pattern of mating and reproduction. In real captive populations, these patterns are often far from the "ideal" described above (Briscoe *et al.* 1992; Lacy 1995). As census size is often limited in captive populations, the next best strategy for preserving variation (after immortality) is to manipulate mating and reproduction to maximize the ratio of the effective to the census size (N_e/N). Effective population size is decreased (drift is increased) in a predictable manner with increasing age structure, imbalance in sex ratio, variance in reproductive success, and variance in population size (Crow and Kimura 1970; Lande and Barrowclough 1987).

Preserving Variation: The "General Strategy"

A general strategy for retaining heterozygosity comes directly from equations relating the census size to the effective population size (Crow and Kimura 1970; Nei et al. 1975; Lande and Barrowclough 1987). Its principles are well understood, as follows: maximize the number of founders, grow as rapidly as possible to carrying capacity, maximize the number of breeders per generation, equalize family sizes, equalize the sex ratio of breeders, and reduce fluctuations in population size (Foose 1983). Adaptation to captivity, which can be extensive (Frankham and Loebel 1992), is also minimized with this strategy; equalizing family sizes reduces adaptation by limiting selection to within families (Haldane 1924; Allendorf 1993).

This general strategy of maximizing N_e/N is actually accomplished through careful orchestration of who mates with whom and how many offspring they have. Wright's methods of maximum avoidance of inbreeding and circular breeding are two such orchestrations (Wright 1977a, p.199–204). The first accomplishes the above strategy by delaying inbreeding as long as possible, which maximizes the inbreeding effective size. The greater the number of individuals, the longer the delay (Kimura and Crow 1963a). Circular breeding instead arranges individuals conceptually in a circle and mates each individual to its two respective neighbors. These "linked" half sibs form the next generation, and the process is repeated. Though inbreeding increases initially more rapidly with this strategy, Kimura and Crow (1963a) show that rates of loss are slower over the long term. This latter strategy has been adapted conceptually for management of subpopulations within which individuals cannot be manipulated (Princée 1995).

A major drawback of these two strategies is that populations may come under management only after several generations of "skewed" or highly non-ideal breeding, and the strategies provide no means by which to decide which individuals should be preferentially bred. Since skew in zoo pedigrees is often extreme, compensatory strategies could potentially have large impact (Ballou and Lacy 1995). With a known pedigree, managers should be able to retain the genetic variation of the original population better by favoring animals from underrepresented lineages than by simply assuming the first managed generation is unrelated and applying the general strategy (Ballou and Lacy 1995).

3.2 UTILIZING PEDIGREE INFORMATION

Basic Principles and Early Difficulties

One means for taking pedigree relationships into account can be summed up in a single measure known as "founder genome equivalents" (Lacy 1989). The status of founders' representation in the present population is quantified using two concepts: "founder contribution" (p_i), the expected proportion of the present population's gene pool that has descended from founder i (if there are ten founders, ideally p_i should be $1/10$); and allelic retention r_i, which is the proportion of founder i's alleles that have survived to the present (ideally, 100%). Founder genome equivalents are thus the number of unrelated founders that would be required to provide the heterozygosity observed in the present population if these founders were all equally represented and had lost no alleles. This is given as:

$$\text{founder genome equivalents} \equiv \frac{1}{2(1 - H_t/H_0)} = \frac{1}{\sum_i p_i^2/r_i},$$

(Lacy 1989), where the sum is over all founders. It can be understood intuitively as follows. Loss of heterozygosity in a complicated pedigree can be simulated by representing the network of relationships in a computer program, assigning two unique alleles to each founder at the top of the pedigree (i.e., 100% heterozygosity), and allowing stochastic Mendelian transmission of the alleles through the pedigree (a "gene drop" simulation; MacCluer et al. 1986). Alleles left at the bottom of the pedigree are used to calculate observed heterozygosity. If this is repeated many times and the results are averaged, an estimate of the expected proportion of heterozygosity remaining in the living population is provided. This H_t is put into the standard equation for describing loss of heterozygosity $H_t = H_0[1 - 1/(2N_e)]^t$ with $H_0 = 1$ and t = the number of generations in the pedigree, and is solved for N_e. The result is the founder genome equivalents for that population. The difference between the actual number of founders and this measure represents the *cumulative* loss of H in the pedigree, in the same way that the difference between census and effective population sizes represents the *per generation* loss in H (Lacy 1995).

Skewed retention (r_i), cannot be fixed with *any* breeding strategy: once alleles are lost, they are lost forever. However, what *can* be fixed is the prominence these remaining alleles have in the extant population. For example, imagine a population descended from three founders. Imagine further that 90% of founder #1's alleles have been lost through skewed breeding, but only 50% of founder #2's and #3's have been lost. If contribution to the present population is equal (i.e., if 1/3 of extant alleles come from each founder), #1 will be overrepresented: the space "allotted" for 100% of #1's alleles is taken up by only 10% of them, while the space "allotted" to #2's and #3's is taken up by 50%; the remaining alleles of #1 are too common relative to what they were in the original population. Founder genome equivalents are thus maximized when each founder contribution is proportional to its retention. This is known as the "target founder contribution"(Ballou and Foose 1996).

Using Pedigree Information: Later Improvements

While target founder contributions are a useful concept for managing simple pedigrees, and have been demonstrated to retain heterozygosity better than random mating (Loebel et $al.$ 1992), they are difficult to apply to complex pedigrees, in which individuals may contain both underrepresented and overrepresented alleles and in which r_i can only be determined by simulation (Thomas 1990). Therefore, several measures of genetic importance have been developed to rank individuals. One of these, the "founder importance coefficient," is defined for animal j as $\sum_i(p_i \times p_{ij})$, where the sum is over all i founders, and p_{ij} is the contribution of founder i to individual j (Ballou and Lacy 1995). Preferentially mating individuals with low founder importance coefficient scores equalizes founder representation; however, it does not take into account allelic retention. Another measure, "genome uniqueness," weights individuals preferentially by how likely they are to carry an allele unique within the population (MacCluer et $al.$ 1986). Strategies using this measure have been used to explore the consequences of maximizing retention of allelic diversity rather than retention of heterozygosity (Allendorf 1986; Thompson 1986). Although allelic diversity appears to be important for the long-term limit to directional selection (James 1971), genome uniqueness is flawed as a measure in two main ways: it is highly computationally intensive, and it includes no way of prioritizing individuals carrying alleles that are at high risk of loss but are not unique (Ballou and Lacy 1995).

Theoretical Advantages of Mean Kinship

Mean kinship values provide yet another way to rank individuals according to genetic importance. The "kinship coefficient" between any pair of individuals is the probability that alleles drawn randomly at the same locus in both individuals are identical by descent (Falconer and MacKay 1996). This coefficient is what the inbreeding coefficient of hypothetical offspring of this pair would be, and it can be calculated for every pair of individuals in the extant population using the known pedigree and an additive relationship matrix (Ballou 1983). The "mean kinship" (MK) for an individual is simply the average of all of its associated kinship coefficients, and it describes how related that individual is on average to everyone else in the current population.

The average MK of a whole population can be expressed in terms of the quantities p_i and r_i, defined above:

$$\text{avg. MK} = (1/2) \sum_i (p_i^2/r_i)$$

(Ballou and Lacy 1995). This is true because the average MK for a population is what the inbreeding coefficient of the next generation would be if mating were random; that is, the probability of identity by descent across the population. This probability is simply $1/(2N_e)$, where N_e in this case is the *founder genome equivalents* defined above. Therefore, the MK coefficient for each individual takes both p_i and r_i into account and improves on earlier measures precisely by accomplishing this without requiring their separate calculation.

The exact use of MK values in particular management situations depends heavily on the life history and demographic characteristics of the species and population to be managed. These constraints can be incorporated using a measure closely related to MK, the *kinship value* (Ballou and Lacy 1995). This is also the mean of kinship coefficients, but each component coefficient is weighted by the appropriate age-dependent reproductive value of population members. Broadly speaking, however, MK is used by calculating its value for all reproductive-age individuals in the population, and by ranking individuals from lowest to highest MK score. Those with low scores carry on average underrepresented lineages, while those with high scores have overrepresented lineages. The lowest-ranked male is then paired with the lowest-ranked female (so as not to mix over and underrepresented lineages). If individuals are able to be paired with multiple mates over their lifetimes, they are assumed to produce one offspring. These hypothetical offspring are added to the population with their parents, and MKs are recalculated for the entire population. Individuals are reranked, and again the lowest are paired (Lacy 1995).

This process is continued until carrying capacity is filled to generate the set of matings and number of offspring from each needed to maximize retention of genetic diversity (which is the same as equalizing MK scores). Animals are then bred aiming towards this ideal; lists of matings and offspring are revised as actual offspring are born and pairings unexpectedly fail or breeders die. When the currently possible ideal is reached, the general strategy described above (equalize family size, etc.) is followed. Due to concerns about inbreeding depression (see below), pairings with a kinship coefficient greater than the population mean inbreeding coefficient are generally avoided (Ballou and Lacy 1995).

3.3 EMPIRICAL EVIDENCE

Direct Evidence Supporting Mean Kinship

A considerable number of studies have experimentally confirmed that components of the general breeding strategy do indeed retard the loss of heterozygosity relative to random mating (Briscoe *et al.* 1992; Frankham and Loebel 1992; Loebel *et al.* 1992; Spielman and Frankham 1992; Borlase *et al.* 1993; Woodworth *et al.* 1994; Backus *et al.* 1995). In addition, two studies have specifically pitted MK versus other weighting methods. The first, cited extensively above (Ballou and Lacy 1995), used computer simulation to test the performance of MK versus maximum avoidance of inbreeding, genome uniqueness, random mating, and founder importance coefficient in the retention of heterozygosity over twenty generations. Five different starting pedigrees were used, which differed in the extent of skew in founder representation. Each treatment consisted of fifty simulated populations of thirty individuals. MK outperformed all other strategies in both retention of heterozygosity and retention of allelic diversity, while also achieving the lowest average level of inbreeding in the population.

The second study directly concerned with evaluation of MK, experimental work using replicate populations of *Drosophila melanogaster*, with about thirty replicate populations of ten individuals each, found that MK retained heterozygosity better than random mating (Montgomery *et al.* 1997). Interestingly, this study also examined the general strategy, and found no significant difference in heterozygosity retained between it and random mating. This suggests that when skew is high (as it is in many zoo populations), far more diversity is lost by not utilizing pedigree information than by not changing breeding practices *per se*.

Problems with Previous Empirical Tests

The primary weakness of these studies is the emphasis placed on determining whether breeding practices retain heterozygosity as expected, rather than exploring whether, *having achieved that*, this greater retention achieves the desired effect of retaining adaptive variation. The above-listed studies using *Drosophila* attempt to achieve this ultimate goal by measuring changes in fitness correlates. However, the lack of quantitative theoretical predictions for loss in mean fitness makes mixed results difficult to interpret.

Inbreeding depression, the decline in the mean value of any character (but particularly fitness) upon mating among relatives is a long-recognized phenomenon affecting many traits, in a wide variety of both wild and captive species, to highly variable degrees (Darwin 1876; Wright 1977b; Ballou and Ralls 1982; Ralls and Ballou 1982, 1983; Nevo et al. 1984; Ralls et al. 1988; Jiménez et al. 1994; Keller et al. 1994). Theories to explain this phenomenon divide broadly into two categories; the first focuses on deleterious recessive or partially recessive alleles (Davenport 1908), which persist in large outbreeding populations but become fixed by chance in small populations (leading to population-level deterioration) and are then also more likely to become homozygous in individuals (leading to individual-level depression). The second focuses on dominance, positing either overall directional dominance for fitness-related characters (that is, with alleles increasing the value of a trait tending to be dominant to decreasing alleles; Fisher 1958; Mather 1955), or generalized overdominance (East 1908), resulting from a supposed greater homeostatic ability gained from multiple protein forms (Mitton and Grant 1984; Mitton 1997). While the two theories are not mutually exclusive, the deleterious recessive model appears to be more generally supported (Frankel 1983; Charlesworth and Charlesworth 1987; Barrett and Charlesworth 1991; Crow 1993; Roff 1997).

In either case, unknown initial distributions of directional dominance and deleterious recessives in populations make prediction of the magnitude of inbreeding depression difficult. This is reflected in the variability of experimental results: for example, Borlase et al. 1993 and Woodworth et al. 1994 found that on average the optimal breeding strategy tested retained higher competitive fitness, but Loebel et al. 1992 and Montgomery et al. 1997 found no effect. Similarly, correlational data are suggestive but solid conclusions remain elusive (Soulé 1980; Nevo et al. 1984; Allendorf and Leary 1986; Caughley 1994; Hedrick 1996; Mitton 1997). Studies that attempt to correlate heterozygosity with mean fitness are hampered by the large sampling variance inherent in estimating genome-wide heterozygosity with allozyme data from relatively few loci (Chakraborty 1981, 1987; Archie 1985; Leberg 1992), the difficulty of measuring fitness, the environmental dependence of fitness, and the fact that other factors such as selection, migration, and mutation may influence heterozygosity and fitness in natural populations much more strongly than they do each other (Hedrick and Miller 1992; Hedrick 1996).

Since precise predictions of the amount of depression to expect are not possible, a "rule of thumb" has prevailed instead. This rule suggests that inbreeding coefficients should not be allowed to increase more than 1% per generation, and is based on the collective experience of domestic animal breeders (Hedrick and Miller 1992). Thus, while assessments of inbreeding depression are important (as short-term population persistence is certainly jeopardized by fitness declines), they are likely to have highly population-specific answers. Perhaps, more importantly, they ignore the long-term question: does population management for retaining heterozygosity retain adaptive ability?

3.4 INSIGHTS FROM QUANTITATIVE GENETIC THEORY

Adaptive Variation

Predictions about changes in adaptive variation are derived from basic population genetic theory. The ability of a population to respond to selection on a trait (its "adaptive ability") is directly proportional to the heritability of that trait (Fisher 1958). Narrow-sense heritability (h^2) is defined as V_a/V_t, where V_t is the total observed phenotypic variance (environmental variation along with all genetic variation), and V_a is the portion of that variation that can be passed reliably from parent to offspring (Falconer and MacKay 1996). This portion derives primarily from additive gene action, although a portion of variation due to dominance can be passed on in the short term. Other genetic variances include remaining dominance and epistatic variance, which are statistical reflections of the interaction of alleles within and among loci. Because alleles sort independently (subject to linkage limitations), these interactions are not reliably passed on through the generations, and most of the variation they contribute to the phenotype is not included in V_a.

Thus, loss in additive variance implies a decline in heritability, as long as environmental and other genetic variance remains constant. This additive variance is defined as the variance in breeding values of the individuals in the population and is given (in a simple one-locus, 2-alleles model, in an ideal, equilibrium population) by

$$V_a = 2pq[a + d(q - p)],$$

in which p and q are the frequencies of the 2 alleles, and a and d describe the additive and dominance effects of these alleles, respectively (Falconer and MacKay 1996). The important point of this equation is that, if dominance effects are negligible, $(d = 0)$, then $V_a = 2pqa$, that is, additive variance is proportional to heterozygosity at this locus. When extended to the multilocus, multiallele case, this aspect of the equation does not change (Falconer and MacKay 1996).

Therefore, *if the assumptions of the simple model hold,* loss in heterozygosity should be mirrored by a proportional loss in heritability, at the expected rate of $1/(2N_e)$ per generation.

Heritability, a composite measure of genetic variation, has allowed successful prediction of change in wild populations over a short time (e.g., Grant and Grant 1993, 1995). If evolutionary potential is defined as the short-term ability to respond to selection, then it is therefore theoretically retained using breeding strategies that slow the loss of heterozygosity. However, evolutionary potential, more broadly, is an ability to respond to selection without going extinct in the process. Thus, it is more than simply high heritability — it is high heritability over an extended period of time. Whether current heritabilities are stable enough to allow longer-term forecasting of evolution of multiple characters is controversial (Lande 1979; Turelli 1988; Shaw *et al.* 1995). However, evidence from vertebrate taxa indicates that long-term adaptive changes do indeed occur along the largest principal component of genetic variance, and are influenced in this direction over fairly long periods of time (Schluter 1996).

It is important to note that other factors may be more important if alternative definitions of evolutionary potential are used. For example, if evolutionary potential is defined instead as the maximum deviation possible from the current mean, then allelic diversity may be a much stronger predictor of such potential than heterozygosity (Robertson 1960; James 1971). Experimental tests have generally supported Robertson's and James's hypothesis. For example, Jones *et al.* (1968) found that total achieved response to selection for bristle number in populations of *Drosophila* increased with the number of founders, and decreased if the populations suffered a one-generation bottleneck (which depletes allelic diversity much more than heterozygosity).

The theoretical superiority of the MK strategy, therefore, rests on the assumption that heritability will decline, according to these classic expectations, more slowly in MK-managed than in control populations. In turn, these expectations are based on assumptions of neutrality, of predictable decline in additive variance, of constant environmental variance, and on the other basic premises of Hardy-Weinberg equilibrium. The questions are to what extent are these assumptions justified, and is the decline in evolutionary potential proportional to the decline in heterozygosity?

"Neutral" Variation?

One basic assumption underlying the classical expectation is that most variation is "neutral," or not under selection. Within captive-breeding

literature, this assumption of neutrality necessary for classical expectations to hold is addressed primarily with theoretical arguments. These
rest on the recognition that the distinction between "adaptive" and
"neutral" variation is somewhat artificial — an allelic form in one environment may be very strongly selected, while it is completely neutral
in another (Vida 1994). Much genetic variation in natural populations
appears to be neutral or nearly neutral (Kimura 1983). Preserving this
variation is important, though, because present neutral variation may
serve as a pre-adaptation to changed conditions in the future (Kimura
and Takahata 1991). These arguments are further bolstered by the observation that, in the population sizes typical of captive populations,
selection will be largely swamped by drift (Lacy *et al.* 1995). Furthermore, captive environments are generally benign, such that selection is
relaxed and many more variants become selectively neutral (Frankham
et al. 1986; Lande 1995). However, agreement on these points is far
from unanimous (Hedrick 1996).

3.5 Indirect Empirical Support

Studies Correlating H with Variability: Individual Level

I am not aware of any studies that have examined the expectation that
losses in H should be proportional to V_a from within a captive-breeding
perspective. Studies outside this perspective, however, are numerous and
fall into two broad categories. The first are correlational studies, which
observe a number of either individuals or populations at a single point
in time and which correlate heterozygosity with another variable (the
second are "longitudinal" studies, which follow the same population or
populations through time. These are addressed below). Individual-level
correlations have received relatively more attention in the form of studies
of fluctuating asymmetry (Mitton 1997). These studies generally indicate that, within a cohort, individuals with higher heterozygosity have
lower asymmetry (i.e., lower variability), perhaps due to greater developmental stability, although this varies greatly among species (Lerner
1954; Mitton 1978; Allendorf and Leary 1986; Palmer and Strobeck 1986;
Wayne *et al.* 1986; D. Houle, pers. comm.). Numerous studies have
further suggested that this lower asymmetry indicates a greater "buffer-
ing" ability in heterozygotes, leading to higher heterozygote growth and
survival, especially during periods of environmental stress or disease
(reviewed in Allendorf and Leary 1986; Mitton 1997). These studies,
of course, cannot address relative changes in additive variance, as this
concept has no meaning at the level of the individual.

Studies Correlating H with Variability: Population Level

Numerous population-level studies correlating heterozygosity and variability in wild species are reviewed by Lynch (1996) and Mitton (1997), though a rigorous statistical analysis has not been attempted to date (Hedges and Olkin 1985; Gurevich and Hedges 1993). Such an analysis may prove valuable for dispelling the common assumption that populations containing less heterozygosity at a given point in time should also contain less adaptive variation (e.g., O'Brien *et al.* 1983; Quattro and Vrijenhoek 1989; Vrijenhoek 1994). A strong correlation is unlikely for several reasons. First, adaptive variation in correlational studies often must be estimated using total phenotypic variation (V_t) as a surrogate measure. The expected positive correlation between H and V_a could easily be obscured by the possibility mentioned above, that more homozygous individuals are more developmentally unstable for purely environmental reasons, which would increase V_e as heterozygosity decreased. Since V_t equals the sum of V_a, V_e, and the remaining nonadditive genetic variation, a negative correlation between heterozygosity and V_e could easily obscure a positive correlation between heterozygosity and V_a when just V_t is measured (Lynch 1996). In addition, the same sampling problems mentioned above plague studies of variability; sample sizes of both loci and populations required for reasonable confidence limits on correlations are typically much larger than those used in studies (Hedrick 1996; Leberg 1996). Finally, the theory outlined above predicts only a correspondence between *rates* of loss of H and V_a, not their *equilibrium levels*. These levels will be differently influenced by disparate initial conditions and population histories (Spitze 1993). Given all of these reasons to expect no correlation between heterozygosity and additive genetic variability at a single time point, it is perhaps unsurprising that Butlin and Gregenza (1998) found no significant relationship between the two in the sixteen species they examined.

"Longitudinal" Studies: Early Work

A second broad category of study which focuses on the relationship between H and V_a is more compelling. These studies follow the same populations through time, allowing relative rates of change to be assessed. Unfortunately, the statistical demands of such studies have limited their application to a small set of well-studied lab organisms. Nonetheless, their results are suggestive.

Early studies include that of Ayala (1965), who found that the rate of increase in fitness correlates in captivity was higher in experimental populations of *Drosophila* derived from more than one natural population. Frankham *et al.* (1968), further found that short-term response to selection was generally lower in populations founded with a small number of flies than with large ones. Since heritability was not broken down into its component parts, however, it is unclear whether this decrease was due to lower additive variance. Hanrahan *et al.* (1973) obtained similar results in mice, but again, no estimate of components was attempted. Hammond (1973) further found that in populations of *Drosophila* founded with one, ten, and fifty pairs and grown to a variety of sizes, those founded with one pair had the lowest heritability for abdominal chaetae, but those established with ten pairs differed little from those founded with fifty pairs, in direct agreement with classic theory. Furthermore, he showed that higher heritability is maintained in populations that are grown as rapidly as possible to carrying capacity — presumably also losing less variance in the process. Finally, Frankham (1980) reported that three unpublished experiments, as well as an additional one of his own, showed that populations of *Drosophila* bottlenecked to one pair of individuals for one generation had about 75% of the heritability for sternopleural bristle numbers as nonbottlenecked lines, as expected from classic theory, if the change was due to loss in additive variance and the magnitude of residual variance was relatively high.

"Longitudinal" Studies: Recent Work

More recently, Briscoe *et al.* (1992) examined populations of *Drosophila* that had been kept in the lab for varying lengths of time, and measured allozyme heterozygosity as well as h^2 and V_a for sterno-pleural bristle numbers. In support of classic theory, all of these measures were negatively correlated with the length of time in captivity. However, since starting measures and breeding structures were not known for any of the populations during this time, the study could not rule out the possibility that populations had by chance started at different levels. Furthermore, the number of populations examined was small ($n = 8$), so that this study can be only strongly suggestive at best. Similarly, Weber (1990) found large reductions in genetic variance for wing shape in *Drosophila* lines that had been in captivity for twenty-two years relative to current wild populations. Again, he had no initial measures of the captive populations with which to compare final measures, and he did not examine additive variance specifically.

Findings in more controlled laboratory studies, both published and unpublished, have been recently reviewed by Fowler and Whitlock (1999). It is clear from this review that the large-scale declines in heritability predicted in classic theory (due to loss of additive variance) are not in-compatible with small-scale increases in phenotypic variation as seen in studies of fluctuating asymmetry. If increased homozygosity leads to increased susceptibility to environmental variance, then this will *also* contribute to a loss of heritability. The authors furthermore point out that when inbreeding is relatively mild (on the level of what might be seen in natural populations), life-history characters appear to be more likely to display increased environmental variance than morphological characters. This review highlights the importance of analyzing losses of heritability both in terms of possible declines in additive variance and in terms of possible increases in environmental variance.

Sample sizes in the studies analyzed were too small on the whole for confident conclusions, however. Therefore, the same authors created fifty-two replicate inbred and outbred lines of *Drosophila* and measured on average ninety parent-offspring families per line in order to estimate declines in additive variance with inbreeding over two generations in various wing-shape characters (Whitlock and Fowler 1999). Although molecular heterozygosity was not directly measured and the number of generations was small, agreement with "classic" theory was strong. On average, additive variation decreased proportionally to the inferred inbreeding coefficient, although there was scatter around this average in individual lines, as expected.

The authors acknowledge that if classic assumptions do not hold, par-ticularly if most genetic variation is due to dominance and/or epistasis, then additive variance can actually increase with inbreeding (Robertson 1952; Goodnight 1987, 1988; Willis and Orr 1993). This "conversion" is most easily imagined in a two-locus model, in which additive-by-additive epistatic variance shifts to purely additive variance if one locus should become fixed, perhaps by drift during a contraction of effective popula-tion size (Carson 1990).

Findings not in conflict with such conversion have in fact been ob-served in laboratory populations of houseflies (e.g., Bryant and Meffert 1995, 1996), flour beetles (Fernández *et al.* 1995; Ruano *et al.* 1996), and fruit flies (López-Fanjul and Villaverde 1989). However, replication in several of these studies is low, and the bottlenecks are of very short duration. As a result, many of these studies cannot statistically distin-guish between true deviations from classical expectations and sampling error around that expectation (Fowler and Whitlock 1999). There is a trend in these studies toward greater deviation from expectation in life-history characters than in morphological ones, which is consistent with greater nonadditive structure in life-history characters (Roff and Mousseau 1987). However, if this is the case, these traits will also be more susceptible to inbreeding depression. If an increase in additive variance of a trait related to fitness is invariably accompanied by inbreeding depression, Willis and Orr (1993, p.956) note that "it seems unlikely that a population following a bottleneck will have a greater evolutionary potential than the ancestral population," since greater variance comes only at the expense of the mean. This is exactly the conclusion of Ruano *et al.* (1996) when they observed an increase in additive variance and a simultaneous sharp decline in the mean of a composite fitness measure in *Tribolium* following inbreeding.

3.6 CONCLUSIONS

The use of genetic techniques in conservation has been criticized in recent years, as too often overemphasized—a high-tech quick fix for problems that are in fact complicated and more often in need of low-tech solutions (Lewontin 1991; Caughley 1994; Snyder *et al.* 1996; Hedrick 1996). While habitat protection and preventative *in-situ* efforts are clearly the preferred routes to conservation, captive breeding and eventual release will be necessary as long as populations are allowed to dwindle to small numbers before action is taken. And as long as breeding is necessary, investigation into its optimal application is crucial.

On the whole, the above evidence indirectly supports the central premise of the mean kinship measure of genetic importance, that loss of additive variance follows classical expectations. Initial empirical work on MK appears promising; the proximate goal of retention of heterozygosity appears to be achieved in populations that are managed by using this measure. Clearly, further study is necessary to determine whether MK, set to be used in the management of many of our most imperiled species, also aids medium-term goals of retention of quantitative variation in actual captive populations; and ultimately, whether it increases the chance of persistence of reintroduced populations in the wild. Answers to these questions will most likely come from a combination of continued study in model systems, ideally expanding beyond the realm of *Drosophila*, and incorporating study of both morphological and life history traits. Also needed is a rigorous experimental approach to actual species reintroductions, to determine whether management with MK can be successfully integrated with species-specific solutions to the considerable social, behavioral, nutritional, developmental, and physiological problems that also face captive populations destined for release.

Acknowledgments

I am grateful for support from Princeton University's Department of Ecology and Evolutionary Biology and the National Science Foundation while completing this work. I thank Lila Fishman, Hope Hollocher, Henry Horn, Lukas Keller, Robert Lacy, Jon Paul Rodríguez, Donald Stratton, Todd Vision, and two anonymous reviewers for comments.

References

Allendorf, F. W. (1986). Genetic drift and the loss of alleles versus heterozygosity. *Zoo Biol.*, *5*, 181–190.

Allendorf, F. W. (1993). Delay of adaptation to captive breeding by equalizing family size. *Conserv. Biol.*, *7(2)*, 416–419.

Allendorf, F. W., and Leary, R. F. (1986). Heterozygosity and fitness in natural populations of animals. In Soulé, M. E. (ed.), *Conservation Biology — The Science of Scarcity and Diversity*, 57–76, Sunderland, Mass.: Sinauer.

Amos, W., and Harwood, J. (1998). Factors affecting levels of genetic diversity in natural populations. *Phil. Trans. Royal Soc. Lond. B.*, *353*, 177–186.

Archie, J. W. (1985). Statistical analysis of heterozygosity data: Independent sample comparisons. *Evolution*, *39(3)*, 623–627.

Arnold, S. J. (1995). Monitoring quantitative genetic variation and evolution in captive populations. In Ballou, J. D., M. Gilpin and T. J. Foose (eds.), *Population Management for Survival and Recovery: Analytical Methods and Strategies in Small Population Conservation*, 295–317, New York: Columbia University Press.

Ayala, F. J. (1965). Evolution of fitness in experimental populations of *Drosophila serrata*. *Science*, *150*, 903–905.

Backus, V. L., Bryant, E. H., Hughes, C. R., and Meffert, L. (1995). Effect of migration or inbreeding followed by selection on low-founder-number populations: Implications for captive breeding programs. *Conserv. Biol.*, *9(5)*, 1216–1224.

Ballou, J. D. (1984). Strategies for maintaining genetic diversity in captive populations through reproductive technology. *Zoo Biol.*, *3*, 311–324.

Ballou, J. D. (1983). Calculating inbreeding coefficients from pedigrees. In Schoenwald-Cox, C. M., Chambers, S. M., MacBryde, F. and Thomas, L. (eds.), *Genetics and Conservation*, 509–520, Menlo Park, CA: Benjamin-Cummings.

Ballou, J. D., and Foose, T. J. (1996). Demographic and genetic management of captive populations. In Kleiman, D. G., M. E. Allen, K. V. Thompson and S. Lumpkin (eds.), *Wild Mammals in Captivity*, 263– 283, Chicago: University of Chicago Press.

Ballou, J. D., and Lacy, R. C. (1995). Identifying genetically important individuals for management of genetic variation in pedigreed populations. In Ballou, J. D., M. Gilpin and T. J. Foose (eds.), *Population Management for Survival and Recovery: Analytical Methods and Strategies in Small Population Conservation*, 76-111, New York: Columbia University Press.

Ballou, J. D., and Ralls, K. (1982). Inbreeding and juvenile mortality in small populations of ungulates: A detailed analysis. *Biol. Conserv.*, *24*, 239–272.

Barrett, S. C. H., and Charlesworth, D. (1991). Effects of a change in the level of inbreeding on the genetic load. *Nature*, *352(8 August)*, 522–524.

Barton, N. H., and Charlesworth, B. (1984). Genetic revolutions, founder effects, and speciation. *Ann. Rev. Ecol. Systematics*, *15*, 133–164.

Beck, B. B., Rapaport, L. G. , Stanley Price, M. R. and Wilson, A. C. (1994). Chapter 13: Reintroduction of captive-born animals. In Olney, P. J. S., G. M. Mace and A. T. C. Feistner (eds.), *Creative Conservation: Interactive Management of Wild and Captive Animals*, 266–286, London: Chapman and Hall.

Borlase, S., C., Loebel, D. A., Frankham, R., Nurthen, R. K., Briscoe, D. A., and Daggard, G. E. (1993). Modeling problems in conservation genetics using captive *Drosophila* populations: Consequences of equalization of family sizes. *Conserv. Biol.*, *7(1)*, 122–131.

Briscoe, D. A., Malpica, J. M., Robertson, A., Smith, G. J., Frankham, R., Banks, R. G., and Barker, J. S. F. (1992). Rapid loss of genetic variation in large captive populations of Drosophila flies: Implications for the genetic management of captive populations. *Conserv. Biol.*, *6(3)*, 416–425.

Bryant, E. H., and Meffert, L. M. (1995). An analysis of selectional response in relation to a population bottleneck. *Evolution*, *49(4)*, 626–634.

Bryant, E. H., and Meffert, L. M. (1996). Nonadditive genetic structuring of morphometric variation in relation to a population bottleneck. *Heredity*, *77*, 168–176.

Butlin, R. K., and Tregenza, T. (1998). Levels of genetic polymorphism: Marker loci versus quantitative traits. *Phil. Trans. Royal Soc. Lond. B*, *353*, 187–198.

Carson, H. L. (1975). The genetics of speciation at the diploid level. *Am. Naturalist*, *109 (965)*, 83–92.

Carson, H. L. (1990). Increased genetic variance after a population bottleneck. *Trends Ecol. Evol.*, *5(7)*, 228–230.

Caughley, G. (1994). Directions in conservation biology. *J. Animal Ecol.*, *63*, 215–244.

Caughley, G., and Gunn, A. (1996). *Conservation biology in theory and practice.* Cambridge, Mass.: Blackwell Science.

Chakraborty, R. (1981). The distribution of the number of heterozygous loci in an individual in natural populations. *Genetics*, *98*, 461–466.

Chakraborty, R. (1987). Biochemical heterozygosity and phenotypic variability of polygenic traits. *Heredity*, *59*, 19–28.

Charlesworth, D., and Charlesworth, B. (1987). Inbreeding depression and its evolutionary consequences. *Ann. Rev. Ecol. Systematics*, *18*, 237–268.

Crow, J. F. (1993). Mutation, mean fitness, and genetic load. In Futuyma, D., and J. Antonovics (eds.), *Oxford Surveys in Evolutionary Biology*, 3–42, Oxford: Oxford University Press.

Crow, J. F., and Kimura, M. (1970). *An Introduction to Population Genetics Theory.* New York: Harper and Row.

Darwin, C. (1876). *The Effects of Self- and Cross-fertilization in the Vegetable Kingdom.* London: J. Murray.

Davenport, C. B. (1908). Degeneration, albinism, and inbreeding. *Science*, *28*, 454–455.

Dobson, A. P., Mace, G. M., Poole, J., and Brett, R. A. (1992). Conservation biology: The ecology and genetics of endangered species. In Berry, R. J., T. J. Crawford and G. M. Hewitt (eds.), *Genes in Ecology*, 405–430, Cambridge: Blackwell Scientific.

East, E. M. (1908). Inbreeding in corn. *Report of the Connecticut Agriculture Experimental Station*, *1907*, 419–429.

Falconer, D. S., and MacKay, T. F. C. (1996). *Introduction to Quantitative Genetics.* Essex, England: Addison Wesley Longman Limited.

Fernández, A., Toro, M. A., and López-Fanjul, C. (1995). The effect of inbreeding on the redistribution of genetic variance of fecundity and viability in *Tribolium castaneum*. *Heredity*, *75*, 376–381.

Fisher, R. A. (1958). *The Genetical Theory of Natural Selection (2nd ed)*. New York: Dover.

Foose, T. J. (1983). The relevance of captive populations to the conservation of biotic diversity. In Schonewald-Cox, C. M., S. M. Chambers, B. MacBryde and W. L. Thomas (eds.), *Genetics and Conservation: A Reference for Managing Wild Animal and Plant Populations*, 374–401, Menlo Park, CA: Benjamin-Cummings.

Fowler, K., and Whitlock, M. C. (1999). The distribution of phenotypic variance with inbreeding. *Genetics*, in press.

Frankel, R. K. (1983). *Heterosis: Reappraisal of Theory and Practice*. New York: Springer-Verlag.

Frankham, R. K. (1980). The founder effect and response to artificial selection in *Drosophila*. In *Selection experiments in laboratory and domestic animals*, 87–90, Harrogate, UK:Commonwealth Agricultural Bureaux, Farnham Royal.

Frankham, R. K., Hemmer, H., Ryder, O. A., Cothran, E. G., Soulé, M. E., Murray, N. D., and Snyder, N. (1986). Selection in captive populations. *Zoo Biol.*, *5(2)*, 127–138.

Frankham, R. K., Jones, L. P., and Barker, J. S. F. (1968). The effects of population size and selection intensity in selection for a quantitative character in Drosophila: I. Short-term response to selection. *Gen. Res. Cambridge*, *12*, 237–248.

Frankham, R. K., and Loebel, D. A. (1992). Modelling problems in conservation genetics using captive *Drosophila* populations: Rapid genetic adaptation to captivity. *Zoo Biol.*, *11*, 333–342.

Franklin, I. R. (1980). Evolutionary change in small populations. In Soulé, M. E., and Wilcox, B. A. (eds.), *Conservation Biology: An Evolutionary and Ecological Perspective*, 135–149, Sunderland, Mass.: Sinauer.

Gilpin, M., and Willis, C. (1991). MHC and captive breeding: a rebuttal. *Conserv. Biol.*, *5(4)*, 554–555.

Goodnight, C. J. (1987). On the effect of founder events on epistatic genetic variance. *Evolution*, *41(1)*, 80–91.

Goodnight, C. J. (1988). Epistasis and the effect of founder events on the additive genetic variance. *Evolution*, *42(3)*, 441–454.

Grant, B. R., and Grant, P. R. (1993). Evolution of Darwin's finches caused by a rare climatic event. *Proc. Royal Soc. London*, *251*, 111–117.

Grant, P. R., and Grant, B. R. (1995). Predicting microevolutionary responses to directional selection on heritable variation. *Evolution*, *49(2)*, 241–251.

Gurevich, J., and Hedges, L. V. (1993). Meta-analysis: Combining the results of independent experiments. In Scheiner, S. M., and J. Gurevich (eds.), *Design and Analysis of Ecological Experiments*, 378–398, New York: Chapman and Hall.

Haldane, J. B. S. (1924). A mathematical theory of natural and artificial selection. *Trans. Camb. Phil. Soc.*, *23*, 19–41.

Haldane, J. B. S., and Jayakar, S. D. (1963). Polymorphism due to selection of varying direction. *J. Genetics*, *58*, 237–242.

Hammond, K. (1973). Population Size, Selection Response and Variation in Quantitative Inheritance. Ph.D. dissertation, Sydney, University of Sydney.

Hanrahan, J. P., Eisen, E. J. and Legates, J. E. (1973). Effects of population size and selection intensity on short-term response to selection for post-weaning gain in mice. *Genetics*, *73(March)*: 513–530.

Hardy, G. H. (1908). Mendelian proportions in a mixed population. *Science*, *28(10 July)* 41–50.

Hedges, L. V., and Olkin, I. (1985). *Statistical Methods for Meta-Analysis*. New York: Academic Press.

Hedrick, P. W. (1996). Conservation genetics and molecular techniques: A perspective. In Smith, T. D., and R. K. Wayne (eds.), *Molecular Genetic Approaches in Conservation*, 459–477, New York: Oxford University Press.

Hedrick, P. W., and Miller, P. S. (1992). Conservation genetics: Techniques and fundamentals. *Ecol. Appl., 2(1)*, 30–46.

Hedrick, P. W., and Miller, P. S. (1994). Rare alleles, MHC, and captive breeding. In Loeschcke, V., J. Tomiuk and S. K. Jain (eds.), *Conservation Genetics*, 187–201, Basel, Switzerland: Birkhäuser Verlag.

Hollocher, H. (1996). Island hopping in *Drosophila*: Patterns and processes. *Phil. Trans. Royal Soc. Lond., B*, 735–743.

Hughes, A. L. (1991). MHC polymorphism and the design of captive breeding programs. *Conserv. Biol., 5(2)*, 249–251.

James, J. W. (1971). The founder effect and response to artificial selection. *Genet. Res., 16*, 241–250.

Jiménez, J. A., Hughes, K. A., Alaks, G., Graham, L., and Lacy, R. C. (1994). An experimental study of inbreeding depression in a natural habitat. *Science, 266(14 October)*, 271–273.

Jones, L. P., Frankham, R. and Barker, J. S. F. (1968). The effects of population size and selection intensity in selection for a quantitative character in Drosophila: II. Long-term response to selection. *Gen. Res. Cambridge, 12*, 249–266.

Keller, L. F., Arcese, P., Smith, J. N. M., Hochachka, w. M., and Stearns, S. C. (1994). Selection against inbred song sparrows during a natural population bottleneck. *Nature, 372(24 November)*, 356–357.

Kimura, M. (1983). *The Neutral Theory of Molecular Evolution*. Cambridge: Cambridge University Press.

Kimura, M., and Crow, J. (1963a). On the maximum avoidance of inbreeding. *Genetical Research, 4(3)*, 399–415.

Kimura, M., and Crow, J. F. (1963b). The measurement of effective population number. *Evolution, 17(3)*, 279–288.

Kimura, M., and Crow, J. F. (1964). The number of alleles that can be maintained in a finite population. *Genetics, 49*, 725–738.

Kimura, M., and Takahata, N. (1991). *New aspects of the genetics of molecular evolution*. New York: Springer-Verlag.

Kleiman, D. G., Allen, M. E., Thompson, K. V., and Lumpkin, S. (1996). Wild Mammals in Captivity. Chicago, University of Chicago Press.

Lacy, R. (1989). Analysis of founder representation in pedigrees: Founder equivalents and founder genome equivalents. *Zoo Biol., 8*, 111–124.

Lacy, R. C. (1995). Clarification of genetic terms and their use in the management of captive populations. *Zoo Biol., 14*, 565–578.

Lacy, R. C., Ballou, J. D., Princée, F., Starfield, A., and Thompson, E. (1995). Pedigree analysis for population management. In Ballou, J. D., M. Gilpin and T. J. Foose (eds.), *Population Management for Survival and Recovery: Analytical Methods and Strategies in Small Population Conservation*, 57–75, New York: Columbia University Press.

Lande, R. (1976). The maintenance of genetic variability by mutation in a polygenic character with linked loci. *Genetical Research, 26*, 221–235.

Lande, R. (1979). Quantitative genetic analysis of multivariate evolution applied to brain: Body size allometry. *Evolution, 33(1)*, 402–416.

Lande, R. (1988). Genetics and demography in biological conservation. *Science, 241(16 September)*, 1455–1460.

Lande, R. (1995). Breeding plans for small populations based on the dynamics of quantitative genetic variance. In Ballou, J. D., M. Gilpin and T. J. Foose (eds.), *Population Management for Survival and Recovery: Analytical Methods and Strategies in Small Population Conservation*, 318–340, New York: Columbia University Press.

Lande, R., and Barrowclough, G. F. (1987). Effective population size, genetic variation, and their use in population management. In Soulé, M. E. (ed.), *Viable Populations for Conservation*, 87–123, Cambridge: Cambridge University Press.

Lande, R., and Shannon, S. (1996). The role of genetic variation in adaptation and population persistence in a changing environment. *Evolution, 50(1)*, 434–437.

Leberg, P. L. (1992). Effects of population bottlenecks on genetic diversity as measured by allozyme electrophoresis. *Evolution, 46(2)*, 477–494.

Leberg, P. L. (1996). Applications of allozyme electrophoresis in conservation biology. In Smith, T. B., and R. K. Wayne (eds.), *Molecular Approaches in Conservation*, 87–103, New York: Oxford University Press.

Lerner, I. M. (1954). *Genetic Homeostasis*. Edinburgh, Scotland: Oliver and Boyd.

Levene, H. (1953). Genetic equilibrium when more than one niche is available. *Journal of Genetics, 55(September-October)*, 511–524.

Lewontin, R. C. (1991). Twenty-five years ago in genetics: Electrophoresis in the development of evolutionary genetics: Milestone or millstone? *Genetics, 128*, 657–662.

Loebel, D. A., Nurthen, R. K., Frankham, R., Briscoe, D. A., and Craven, D. (1992). Modelling problems in conservation genetics using captive *Drosophila* populations: Consequences of equalizing founder representation. *Zoo Biol., 11*, 319–332.

López-Fanjul, C., and Villaverde, A. (1989). Inbreeding increases genetic variation for viability in *Drosophila melanogaster*. *Evolution, 43(8)*, 1800–1804.

Lynch, M. (1996). A quantitative-genetic perspective on conservation issues. In Avise, J. C. (ed.), *Conservation Genetics: Case Histories from Nature*, 471–501, New York: Chapman Hall.

MacCluer, J. W., VandeBerg, J. L., Read, B., and Ryder, O. (1986). Pedigree analysis by computer simulation. *Zoo Biol., 5*, 147–160.

Magin, C. D., Johnson, T. H., Groombridge, B., Jenkins, M., and Smith, H. (1994). Species extinctions, endangerment, and captive breeding. In Olney, P. J. S., G. M. Mace and A. T. C. Feistner (eds.), *Creative conservation: Interactive management of wild and captive animals*, 3–31, London: Chapman and Hall.

Mather, K. (1955). The genetical basis of heterosis. *Proc. Roy. Soc. Lond., B144*, 143–150.

May, R. M., Lawton, J. H., and Stork, N. E. (1995). Assessing extinction rates. In Lawton, J. H., and R. M. May (eds.), *Extinction Rates*, 1–24, New York: Oxford.

Mayr, E. (1963). The genetics of speciation. In Mayr, E. (ed.), *Animal Species and Evolution*, 516–555, Cambridge: Harvard University Press.

Mitton, J. B. (1978). Relationship between heterozygosity for enzyme loci and variation of morphological characters in natural populations. *Nature, 273*, 661–662.

Mitton, J. B. (1997). *Selection in natural populations*. Oxford: Oxford University Press.

Mitton, J. B., and Grant, M. C. (1984). Associations among protein heterozygosity, growth rate, and developmental homeostasis. *Ann. Rev. Ecol. Systematics, 15*, 479–499.

Montgomery, M. E., Ballou, J., Nurthen, R. K., England, P. R., Briscoe, D. A., and Frankham, R. (1997). Minimizing kinship in captive breeding programs. *Zoo Biol., 16(5)*, 377–389.

Nei, M. (1978). Estimation of average heterozygosity and genetic distance from a small number of individuals. *Genetics, 89(July)*, 583–590.

Nei, M., Maruyama, T., and Chakraborty, R. (1975). The bottleneck effect and genetic variability in populations. *Evolution, 29(1)*, 1–10.

Nevo, E., Beiles, A., and Ben-Shlomo, R. (1984). The evolutionary significance of genetic diversity: Ecological, demographic, and life-history correlates. *Evolutionary Dynamics of Genetic Diversity (Lecture Notes in Biomathematics), 53*, 13–213.

O'Brien, S. J., Wildt, D. E., Goldman, D., Merril, C. R., and Bush, M. (1983). The cheetah is depauperate in genetic variation. *Science, 221(29 July)*, 459–462.

Palmer, A. R., and Strobeck, C. (1986). Fluctuating aysmmetry: Measurement, analysis, patterns. *Ann. Rev. of Ecol. and Systematics, 17*, 391–421.

Princée, F. P. G. (1995). Overcoming the constraints of social structure and incomplete pedigree data through low-intensity genetic management. In Ballou, J. D., M. Gilpin and T. J. Foose (eds.), *Population Management for Survival and Recovery: Analytical Methods and Strategies in Small Population Conservation*, 124–154, New York: Columbia University Press.

Provine, W. B. (1989). Founder effects and genetic revolutions in microevolution and speciation: An historical perspective. In L.V. Giddings, K. Y. Kaneshiro, and W. W. Anderson, ed., *Genetics, Speciation, and the Founder Principle*. Oxford: Oxford University Press.

Quattro, J. M., and Vrijenhoek, R. C. (1989). Fitness differences among remnant populations of the endangered Sonoran topminnow. *Science, 245*, 976–978.

Ralls, K., and Ballou, J. (1982). Effects of inbreeding on infant mortality in captive primates. *Internatl. J. Primatology, 3(4)*, 491–505.

Ralls, K., and Ballou, J. (1983). Extinctions: Lessons from zoos. In Schonewald-Cox, C. M., S. Chambers, B. MacBryde and W. Thomas (eds.), *Genetics and conservation: A reference for managing wild animal and plant populations*, 164–184, London: Benjamin/Cummings.

Ralls, K., Ballou, J. D., and Templeton, A. (1988). Estimates of lethal equivalents and the cost of inbreeding in mammals. *Conserv. Biol., 2(2)*, 185–193.

Ralls, K., and Meadows, R. (1993). Breeding like flies. *Nature, 361(25 February)*, 689–690.

Rice, W. R. (1993). Laboratory experiments on speciation: What have we learned in 40 years? *Evolution, 47(6)*, 1637–1653.

Robertson, A. (1952). The effect of inbreeding on variation due to recessive genes. *Genetics, 37(March)*, 189–207.

Robertson, A. (1960). A theory of limits in artificial selection. *Proc. Roy. Soc. Lond., 153B*, 234–249.

Roff, D. A. (1997). *Evolutionary Quantitative Genetics*. New York: Chapman and Hall.

Roff, D. A., and Mousseau, T. A. (1987). Quantitative genetics and fitness: Lessons from *Drosophila*. *Heredity, 58*, 103–118.

Rose, M. R. (1982). Antagonistic pleiotropy, dominance, and genetic variation. *Heredity, 48(1)*, 63–78.

Ruano, R. G., Silvela, L. S., Lopez-Fanjul, C., and Toro, M. A. (1996). Changes in the additive variance of a fitness-related trait with inbreeding in *Tribolium castaneum*. *J. Animal Breed. Genet., 113(2)*, 93–97.

Schluter, D. (1996). Adaptive radiation along genetic lines of least resistance. *Evolution, 50(5)*, 1766–1774.

Shaw, F. H., Shaw, R. G., Wilkinson, G. S., and Turelli, M. (1995). Changes in genetic variances and covariances: G whiz! *Evolution, 49(6)*, 1260–1267.

Shoemaker, A., and Flesness, N. (1996). Records, Studbooks; and ISIS inventories. In Kleiman, D. G., M. E. Allen, K. V. Thompson and S. Lumpkin (eds.), *Wild Mammals in Captivity*, 600–603, Chicago: University of Chicago Press.

Simpson, G. G. (1953). *The Major Features of Evolution*. New York: Columbia University Press.

Snyder, N. F. R., Derrickson, S.R., Beissinger, S. R., Wiley, J. W., Smith, T. B., Toone, W. D., and Miller, B. (1996). Limitations of captive breeding in endangered species recovery. *Conserv. Biol.*, *10(2)*, 338–348.

Soulé, M. E. (1980). Thresholds for survival: Maintaining fitness and evolutionary potential. In Soulé, M. E., and Wilcox, B. A. (eds.), *Conservation Biology: An Evolutionary and Ecological Perspective*, 151-169, Sunderland, Mass.: Sinauer.

Spielman, D., and Frankham, R. (1992). Modelling problems in conservation genetics using captive *Drosophila* populations: Improvement of reproductive fitness due to immigration of one individual into small partially inbred populations. *Zoo Biol.*, *11*, 343–351.

Spitze, K. (1993). Population structure in *Daphnia obtusa*: Quantitative genetic and allozymic variation. *Genetics*, *135(October)*, 367–374.

Tauber, C. A., and Tauber, M. J. (1989). Sympatric speciation in insects: Perception and perspective. In Otte, D., and J. A. Endler (eds.), *Speciation and its consequences*, 307–344, Sunderland, Mass.: Sinauer.

Templeton, A. R. (1989). The meaning of species and speciation: A genetic perspective. In Otte, D., and J. A. Endler (eds.), *Speciation and Its Consequences*, 3–27, Sunderland, Mass.: Sinauer.

Thomas, A. (1990). Comparison of an exact and a simulation method for calculating gene extinction probabilities in pedigrees. *Zoo Biol.*, *9*, 259–274.

Thompson, E. A. (1986). Ancestry of alleles and extinction of genes in populations with defined pedigrees. *Zoo Biol.*, *5*, 161–170.

Thompson, E. A. (1995). Genetic importance and genomic descent. In Ballou, J. D., M. Gilpin and T. J. Foose (eds.), *Population Management for Survival and Recovery: Analytical Methods and Strategies in Small Population Conservation*, 112–123, New York: Columbia University Press.

Turelli, M. (1988). Phenotypic evolution, constant covariances, and the maintenance of additive variance. *Evolution*, *42(6)*, 1342–1348.

Vida, G. (1994). Global issues of genetic diversity. In Loeschcke, V., J. Tomiuk and S. K. Jain (eds.), *Conservation Genetics*, 9–19, Boston: Birkhäuser.

Vrijenhoek, R. C. (1994). Genetic diversity and fitness in small populations. In Loeschcke, V., J. Tomiuk and S. K. Jain (eds.), *Conservation Genetics*, 38–53, Boston: Birkhäuser.

Vrijenhoek, R. C., and Leberg, P. L. (1991). Let's not throw out the baby with the bathwater: A comment on management for MHC diversity in captive populations. *Conserv. Biol.*, *5(2)*, 252–254.

Wayne, R. K., Forman, L., Newman, A. K., Simonson, J. M., and O'Brien, S. J. (1986). Genetic monitors of zoo populations: Morphological and electrophoretic assays. *Zoo Biol.*, *5*, 215–232.

Weber, K. E. (1990). Selection on wing allometry in Drosophila melanogaster. *Genetics*, *126*, 975–989.

Weinberg, W. (1963). Über den Nachweis der Vererbung beim Menschen (On the demonstration of heredity in man, 1908). In Boyer, S. H. (ed.), *Papers on Human Genetics*. Englewood Cliffs, NJ: Prentice-Hall.

Whitlock, M. C., and Fowler, K. (1999). The changes in genetic and environmental variance with inbreeding in *Drosophila melanogaster*. *Genetics*, in press.

Willis, J. H., and Orr, H. A. (1993). Increased heritable variation following population bottlenecks: The role of dominance. *Evolution*, *47(3)*, 949–957.

Woodruff, D. S. (1989). The problems of conserving genes and species. In Western, D., and M. C. Pearl (eds.), *Conservation for the twenty-first century*, 76–88, New York: Oxford University Press.

Woodworth, L. M., Montgomery, M. E., Nurthen, R. K., Briscoe, D. A., and Frankham, R. (1994). Modelling problems in conservation genetics using *Drosophila*: Consequences of fluctuating population sizes. *Mol. Ecol.*, *3*, 393–399.

Wright, S. (1931). Evolution in Mendelian populations. *Genetics*, *16(March)*, 97–159.

Wright, S. (1977a). *Evolution and the Genetics of Populations: Volume 2, The Theory of Gene Frequencies*. Chicago: University of Chicago Press.

Wright, S. (1977b). *Evolution and the Genetics of Populations: Volume 3, Experimental Results and Evolutionary Deductions*. Chicago: University of Chicago Press.

4

Two Problems with the Measurement of Genetic Diversity and Genetic Distance

WILLIAM AMOS

SUMMARY. Genetic analysis is often considered a central part of conservation biology. Much of this perception is fully justified. However, there are several circumstances in which genetic analysis is often applied uncritically. This chapter examines two cases. It first questions the widespread use of genetic variability as a diagnostic tool for measuring the evolutionary health of a species. Although population decline undoubtedly carries with it genetic threats, current methods lack power to assess these threats, and several classic studies appear less clear-cut than previously supposed. In a more general sense, it remains unclear what exactly constitutes the "right" amount of diversity. Second, I examine the pitfalls inherent in assuming genetic markers to be passive, well-behaved tools. It seems that microsatellites, arguably the most important class of genetic markers, may have a life of their own. Understanding the rules by which markers evolve will allow us to avoid problems and harness as yet unrecognised potential. As an example, I explore how a putative link between evolutionary rate and heterozygosity may be used to infer historical patterns of population growth and decline.

INTRODUCTION

Recent advances in molecular genetic techniques have revolutionised many areas of biology, not least in behavioural ecology and taxonomy. Here, methods such as DNA fingerprinting and the direct sequencing of mitochondrial DNA now allow us to answer a range of exciting questions that were previously considered either difficult or intractable. One branch of biology in which the impact has been particularly strong is in conservation biology.

The range of molecular genetic techniques being applied appears at first sight to be dauntingly diverse. However, they can all be united by a single common theme. All techniques are based ultimately on the use of genetic similarities and differences to measure relatedness. As such, the gamut of methodologies can be placed on a single scale based only on the evolutionary scale over which they are used. Finest resolution is afforded at the level of the family by techniques such as DNA finger-printing (Jeffreys *et al.* 1985a; Packer *et al.* 1991). Relatedness between populations is used to infer patterns of gene flow and metapopulation structure (Baker *et al.* 1990). Above the level of the species, relatedness between species is used in taxonomy to recover phylogenetic information (Hillis and Moritz 1990).

Although molecular genetic techniques are undoubtedly powerful, we are not yet at the stage at which the results gained can be accepted blindly. Much still remains to be learned both about how DNA sequences evolve and about how the patterns of variability being measured relate to higher level characters such as individual fitness and population dynamics. Here, I examine two aspects of the role of genetics in conservation. First, I discuss the relationship between population bottlenecks, levels of genetic variability, and the genetic problems faced by small populations in an attempt to learn how practical it is to assess the impact of human activities on genetic viability of natural populations. Second, I present evidence about how one of the key classes of genetic markers, microsatellites, may evolve in an unexpected fashion that is likely to confound their use for measuring genetic distance.

4.1 POPULATION BOTTLENECKS AND GENETIC VARIABILITY

Loss of Variability versus Inbreeding Depression

In the literature it is common to find statements that, either explicitly or implicitly, employ the following train of logic. A population that is either currently small, or has in the past suffered a population bottleneck, will have lost variability, and this will lead to inbreeding depression and loss of fitness. Consequently, it is often assumed that by measuring current levels of heterozygosity, the genetic "health" of a population can be measured.

Although superficially reasonable, a closer inspection reveals that this train of logic confounds two distinct concepts. First, there is the threat posed by loss of genetic variability, a danger that is thought to be manifest in terms of lowering a population's ability to adapt to novel threats such as pathogens, competitors, or parasites (Lande and Barrowclough 1987; O'Brien 1994). Quite separate are the detrimental effects of inbreeding depression. Here, population size reduction creates a founder effect in which rare deleterious recessive traits have an opportunity to change in frequency. While many of these recessives are lost, some may drift to high frequency and become expressed in the homozygous state, causing a reduction in individual fitness (Keller *et al.* 1994; Ralls *et al.* 1970; Saccheri *et al.* 1998; Saccheri *et al.* 1996).

There are two important problems which complicate any attempts to infer risk of inbreeding depression from measurements of genetic variability. First, loss of variability and inbreeding depression affect populations at different times and exert their influences over very different timescales. While loss of variability through neutral genetic drift takes place over many generations (Hartl 1988), the effects of inbreeding depression tend to be felt at, or soon after, the population decline (Keller *et al.* 1994). Consequently, immediately after a population crash there may be severe inbreeding depression even though little variability has yet been lost.

Are Population Bottlenecks Detectable?

The second problem is that it is unclear whether biologically plausible population bottlenecks will have been sufficiently severe to be detectable using current techniques. This is particularly so for bottlenecks due to human activities such as habitat degradation and fragmentation and harvesting, since these have, for the most part, had less than a century in which to exert an effect. To illustrate, the rate at which heterozygosity is lost after a population decline depends primarily on the new effective population size (regeneration of variability through new mutations can usually be ignored), and is given by the following equation:

$$(4.1) \qquad H_t = H_0 \left(1 - \frac{1}{2N_e}\right)^t.$$

Here H_t is the heterozygosity after t generations at the new effective population size N_e, in terms of the predecline heterozygosity H_0.

Substituting plausible parameter values reveals that loss of heterozygosity through drift alone occurs extremely slowly. For example, a population of animals with a generation length of five years that was reduced to an effective population size of fifty at the turn of the century would still retain 82% of its initial variability. Calculations such as this can be used to argue that only under exceptional circumstances will a species lose more than 5%–10% of its heterozygosity in less than a century. Consequently, claims of rapid, dramatic impoverishment should be treated with caution. Consider the following well-known examples:

Example 1: The cheetah

The cheetah is a species with legendarily poor variability, high levels of nonfunctional sperm, and low breeding success in zoos (O'Brien 1994). In this review, it is stated: "These studies lend support to the hypothesis that the cheetah's ancestors survived a historic period of extensive inbreeding (a population bottleneck), the modern consequences of which are 90%–99% reduction in measurable allelic variation." Solving Equation 4.1 for the more extreme figure of 99% loss reveals requirements of, for example, either 16 generations at a population size of 2 or 227 generations at a population size of 25. Even a loss of 90% heterozygosity would require over 100 generations at a population size of 25.

Recent population estimates for the African cheetah are of the order of 7,000 (Caro 1994; Macdonald 1984), a figure that is on the way down, not up, yet that is still far too high to be considered a significant bottleneck. The possibility of a historic bottleneck (or bottlenecks) is difficult to rule out, but seems biologically unlikely, requiring a species that is clearly adapted to population sizes measured in thousands to have persisted for very long periods as relict populations of 100 or less. It has been pointed out that certain forms of metapopulation structure can result in effective population sizes that are only a small fraction of the census size (Gilpin 1991; Hedrick 1996). However, this explanation for the cheetah's lack of variability is a long-term equilibrium solution rather than a bottleneck. It is also unclear how such a population structure, which results in such low levels of variability and periodic inbreeding, would evolve given that most species appear to adopt behaviours that avoid inbreeding (Amos *et al.* 1993; Brooker *et al.* 1990).

The case of the cheetah is thus not as straightforward as it first seemed, and we must look beyond any recent population decline caused by human activities for an explanation. In fact, more recent studies have questioned whether genetic variability in the cheetah is low at all, arguing that cats in general are less variable than other mammals (Hedrick 1996). At the same time, others have challenged the notion that cheetahs in the wild are affected adversely by inbreeding depression (Caro and Laurenson 1994; Laurenson *et al.* 1995). A reassessment of cheetah breeding success in captivity suggests that inbreeding depression does affect matings between closely related parents, but that matings between unrelated individuals show similar levels of fertility to other felids (Wielebnowski 1996).

Example 2: The northern elephant seal

If the case of the cheetah fails to hold up to close scrutiny, what of other species with low genetic variability? A second test case is provided by the northern elephant seal, *Mirounga angustirostris*, a species that was hunted for its oil during the nineteenth century (Hoelzel *et al.* 1993). After a final major catch of 153 animals in 1854 (Townsend 1885), sightings fell away, and the species was considered extinct. A few years later, in 1892, Townsend reported eight seals on Guadeloupe Island, of which seven were killed for collections. Despite this merciless attention, the species has recovered, with population size estimates of 350 in 1922, 15,000 in 1960, and 120,000 in 1980 (Stewart *et al.* 1994).

Clearly, Townsend's report that only one individual was left is incorrect: there must have been other animals that went undetected. A crude approximation as to how many more went undetected can be obtained by extrapolating back from the later population estimates. During the interval between 1922 and 1960, an increase equivalent to an average of 10.4% per annum occurred. Between 1960 and 1980, the rate was 14% per year. Taking a figure of 12% per annum and extrapolating back from 350 in 1922, we arrive at an estimate of approximately 100 animals in 1910. The severest part of the population bottleneck was therefore very brief, lasting only twenty-five years. Elephant-seal generation length is difficult to estimate, and almost certainly varies with breeding colony density and other parameters, but it is difficult to imagine that it is less than ten years, suggesting that the bottleneck lasted less than three generations. Using a figure of 5 for the average effective population size during this period, we find that some 73% of nuclear heterozygosity would have been retained. Mitochondrial DNA diversity might have been reduced to an even lower level due to the smaller effective population size of the mtDNA genome (Hedrick 1995; Hoelzel *et al.* 1993; Slade *et al.* 1998).

Once again, a species appears to have less variability than would be predicted by its population history. However, for the elephant seal there is an interesting control species. The Antarctic fur seal, *Arctocephalus gazella*, yielded over one million skins before becoming commercially extinct in 1822 (Bonner 1968). For almost a century occasional animals were killed as and when they were seen, with many careful searches of their traditional breeding beaches failing to reveal a single individual. It is a wonder that the species survived at all. Then, on Bird Island, South Georgia, one and five animals were recorded in 1915 and 1919, respectively (the former was killed!), after which recovery was rapid. Such figures appear to be, if anything, more extreme than those relating to the northern elephant seal, yet current genetic studies reveal high levels of variability (Gemmell *et al.* 1997).

Problems with Measuring Loss of Genetic Diversity

Publication biases

It seems that most biologically plausible bottlenecks will result in a rather small decline in heterozygosity. Therefore, it is interesting to ask whether molecular genetic techniques can detect the levels of depletion that could result. A number of obstacles need to be overcome. First, there is the basic statistical problem of power. To detect a drop in heterozygosity of only 10% requires either large sample sizes (e.g., more than 100) or data from many independent loci. However, in studies of endangered species, the sample sizes of both individuals and markers are likely to be small, and the analysis is likely to be further confounded by inclusion of relatives. Arguably more serious is the fact that poor statistical power will usually be exacerbated by lack of any comparative sample collected before the decline. When such a sample is not available, loss of heterozygosity can be inferred only through comparison with levels of variability measured in related taxa. Given that levels of heterozygosity can vary considerably from species to species for reasons other than population decline, it is clear that small changes in heterozygosity will be lost in background noise. Only exceptionally low levels of heterozygosity will stand out. Ironically, these extreme cases are unlikely to be explicable in terms of drift alone!

Measurement of variability is further complicated by publication biases (Amos and Harwood 1998). In studies of breeding behaviour and population structure involving species that are not thought to be bottlenecked, failure to find variability is seen as an inconvenience. Sometimes the project is abandoned, and sometimes other genetic markers are sought, but seldom is the study published rapidly or in its original form. A good case in question is that of the European badger, which has proved a frustrating stumbling block for several groups attempting to use DNA fingerprinting. No paper has yet emerged, thereby depriving the literature of a good example of a species in which population size is large but variability is inexplicably low. Studies of this kind inevitably bias upwards our picture of what is a "normal" level of variability. Unfortunately, the situation is exacerbated by converse findings. When high levels of variability are found, publication may be rapid because, for example, paternity analysis can be achieved with high confidence and fewer markers. Indeed, there are plenty of examples in which the finding of great variability is itself considered the lead feature worthy of publication (Avise *et al.* 1989; Kipling and Cooke 1990; Wahls *et al.* 1990; Wenink *et al.* 1993).

Now consider parallel studies that focus on species in which a bottleneck is suspected. When "normal" levels of variability are found, there is a natural tendency either not to mention bottlenecks or to hypothesize that numbers never got quite as low as had been suspected. Either way, the point receives no great attention. By contrast, the finding of low variability is usually interpreted as an example of how population decline leads to loss of variability. Few adopt a critical stance and examine whether the "low" levels of variability found are compatible with the population history of the species being studied. This situation is only made worse by the publication bias mentioned in the previous paragraph, because an inflated view of what is a "normal" level of variability will make it more likely that any random study finds a level that can be classified as "low." Preconceptions reinforce preconceptions.

Can we measure loss of diversity?

The conditions under which it is possible to detect a realistic decrease in heterozygosity due to drift alone appear to be rather narrow. On the one hand, the losses tend to be slight, and on the other, the optimal method, comparing samples collected both before and after the decline, is unlikely to be possible in most cases. However, there are a number of ways by which the situation may be improved.

First, heterozygosity is not the only way to measure genetic variability, and other measures such as the total number of alleles can be more sensitive to the effects of population decline (Brookes *et al.* 1997; Leberg 1992), particularly at loci where allelic diversity is extreme. For example, consider a minisatellite locus at which heterozygosity is (effectively) 100%, in a population that is reduced to two individuals and then rebounds. Heterozygosity will have been lost according to Equation 4.1, and will now stand at 75%. However, the number of alleles in the population is now at most four, and will stay that way until new mutations have begun to replenish variability. The process of allele loss is seen most dramatically in the complex, bar-code-like banding patterns of DNA fingerprints, and a good case in question is provided by an elegant study of the Californian Channel Island fox. This species experienced strong founder effects each time it colonized a new island, causing massive loss of fingerprint complexity. Nowadays, each island has essentially a single DNA fingerprint, which characterizes all its members and which distinguishes it from all others (Gilbert *et al.* 1990). Whether or not these profound phenotypic differences are sufficient to raise each population to species status poses an interesting question for biologists to consider.

A second approach that will help provide a better estimate of the degree of genetic loss is to seek comparative samples collected from before the population crash. Although until recently this possibility was usually little more than a pipe dream, the advent of PCR and its refinement for application to problems involving forensic samples and ancient DNA (Hagelberg 1994) means that the dream is now within reach. Museum specimens, including skins, bones, and alcohol-preserved specimens, have all been analyzed successfully, and a long list of species already studied includes the extinct quagga and mammoth, the red wolf, and the loggerhead shrike (Hagelberg *et al.* 1994; Higuchi *et al.* 1984; Mundy *et al.* 1997; Wayne and Jenks 1991).

Third, even though the literature is full of reports that link low population size to low levels of variability, it is clear from population genetic theory that many of these are unlikely to be due to the action of genetic drift alone, and hence there is a need to consider other possibilities. One plausible factor that might contribute is inbreeding, both through inbreeding depression, the selective purging of deleterious recessive alleles (Barrett and Charlesworth 1991; Saccheri *et al.* 1996), and inbreeding avoidance, behaviours that reduce or minimize the risk that close relatives will pair (Gilbert *et al.* 1991). Both these factors have the potential to exacerbate the effects of a bottleneck by acting as a form of filter, reducing or eliminating the representation of some parts of the gene pool in future generations. For example, in very small populations (e.g., less than 30), it is easy to visualize how some individuals could have much reduced or even zero reproductive capability due to a lack of unrelated partners with which to mate.

4.2 MEASURING GENETIC DISTANCE

The ability to determine how closely related one population is to another could be extremely useful in conservation biology. Among other things, it would allow us to assess the significance of newly imposed barriers to gene flow, to manage threatened populations efficiently, and perhaps to help with decisions concerning the genetic augmentation of impoverished populations. In this area, genetic techniques potentially have a lot to offer, but which methodology is best suited?

A Brief History of Molecular Genetic Markers

Around 1960, protein electrophoresis became the first molecular genetic technique to be applied routinely to answer ecological questions and led to the accumulation of a veritable mass of data (Nevo *et al.* 1984).

Since then, the field has witnessed wave upon wave of improvement and change. Protein electrophoresis was exciting, because it was the first technique to allow easy access to extensive genetic variability; however, as more and more data were accumulated, its inherent limitations began to become more apparent. Most of the loci available were metabolic enzymes that carried useful, but rather similar, levels of variability. As understanding of genomes improved, it became clear that this one class of proteins was only the very tip of the variability iceberg, and that far greater potential lay with the study of DNA itself (Majerus *et al.* 1996).

Around 1970, the discovery of restriction enzymes, a class of proteins that cleaved DNA at highly specific sites, opened the door to variability at the level of the DNA molecule itself. To begin with, people used anonymous markers, random pieces of cloned DNA, to generate restriction-fragment-length polymorphisms. Later, there was a move towards the study of specific DNA sequences with particular, and useful, characteristics. Of prime importance here was the rapidly evolving, maternally transmitted mitochondrial genome (Wilson *et al.* 1985). Mitochondrial DNA soon became established as the central source of polymorphic markers for the study of natural populations (Moritz 1994).

Then, in 1985, the application of molecular techniques received an enormous boost with the discovery by Professor Alec Jeffreys of minisatellites and the technique of DNA fingerprinting (Jeffreys *et al.* 1985a). Here was a technique that caught the imagination, allowing identification of individuals (Jeffreys *et al.* 1985b), positive parentage testing, and the measurement of relatedness (Amos *et al.* 1991; Gilbert *et al.* 1991; Packer *et al.* 1991). The result was predictable, a massive wave of new interest and new projects (Burke 1989). However, DNA fingerprinting is technically demanding and methodologically cumbersome to apply to large data sets (Amos and Pemberton 1992). Thus, after a brief period during which exciting results came in equal measure with widespread frustration, more accessible alternatives began to appear. Of these, the latest vogue was microsatellite analysis (Bruford and Wayne 1993).

Microsatellites are short regions of DNA sequence in which a motif of only 2–5 bases is repeated, head to tail, a number of times. These regions are unstable, suffering frequent gain and loss of repeat motifs, a process that causes extensive length polymorphism (Schlötterer and Tautz 1992; Tautz 1989). The advantages of microsatellite polymorphisms over previous genetic markers are well documented. Levels of polymorphism are high and predominantly selectively neutral. Microsatellites are abundant in the genomes of all higher organisms, making them accessible. Being short, variability can be visualized by means of the polymerase chain reaction (PCR), allowing even small and degraded samples to be typed (Jeffreys *et al.* 1992). While we still have much to learn about what DNA sequences genomes contain, we probably know enough to be confident that no other sequences share this combination of ubiquitous abundance, polymorphism, and apparent neutrality, causing one to wonder whether, more than any of its predecessor techniques, microsatellite analysis seems to be here to stay.

Microsatellites are extremely versatile markers, being useful for assessing individual identity, parentage, and relatedness (Amos *et al.* 1993; Dow and Ashley 1996; Queller *et al.* 1993), population structure (Paetkau *et al.* 1995; Valsecchi *et al.* 1997) and even phylogenetic relationships (Bowcock *et al.* 1994). Thus, while mitochondrial DNA remains a key source of matrilineal markers (Moritz 1994), microsatellite analysis is now the method of choice for studies of nuclear gene flow. However, it is important that we do not accept these benefits uncritically. In the following section, I consider how recent findings show how little we still know about how microsatellites evolve. On the one hand this new picture places a large question mark against the way genetic distances are calculated, but the silver lining behind the cloud may be a source of new information with which we may be able to elucidate historical population trends.

Microsatellite Evolution and Genetic Distance

The apparent simplicity of microsatellite evolution has proved to be a powerful lure for theoreticians, creating a large and burgeoning body of literature based on both analytical and computer simulation approaches. The null model used by the majority of these studies combines a symmetric mutation process (gains in length balancing losses) with neutral genetic drift (Goldstein *et al.* 1995a; Slatkin 1995). Mutations are assumed to involve either the gain or the loss of a single repeat unit, the so-called stepwise mutation model, though variations including a two-phase model (Di Rienzo *et al.* 1994), which allows occasional larger jumps, and the infinite alleles model, in which any allele can mutate to any other, have also been examined. Long microsatellites are rare, but since we know nothing about why this is so, most models have been forced to adopt the pragmatic approach of either leaving length unconstrained (Shriver *et al.* 1993) or invoking a reflecting length boundary (Nauta and Weissing 1996). (I have yet to come across a study that considers the third possibility, an absorbing boundary, perhaps created by a stability threshold.)

Out of these studies have been born two statistics, developed specifically around the stepwise mutation model. As a measure of population subdivision, Slatkin formulated an analogue of Wright's inbreeding coefficient, F_{st}, which he has called R_{st} (Slatkin 1995). Around the same time, Goldstein and colleagues published a microsatellite genetic distance measure that they christened $(\delta\mu)^2$ (Goldstein and Pollock 1997; Goldstein et al. 1995a; Goldstein et al. 1995b). Subsequent analysis and computer simulations have shown that over short evolutionary timescales, genetic distance can scale linearly with time, that genetic distances will only be reliable when a large number (300) of loci are studied (Takezaki and Nei 1996), and that microsatellites are likely to be poor at resolving deeper evolutionary nodes.

Unfortunately, early attempts to measure genetic distance using microsatellites have thrown up rather confusing results. First, comparing the relative distances between human populations with the distance between humans and chimpanzees, it was found that the human-chimpanzee distance was too small by up to a factor of nearly ten (Deka et al. 1994; Garza et al. 1995; Rubinsztein et al. 1995b). Second, despite the finding of significant differences in allele frequencies, attempts to determine the relationships between human populations have, by and large, yielded equivocal results. Elsewhere, alternative genetic distance measures applied to different oceanic populations of the humpback whale from the same microsatellite data set, revealed profound disagreements between methods (Valsecchi et al. 1997). In some cases the largest genetic distance according to one method was the smallest according to another. These apparent problems suggest a need to look more critically at the assumptions on which the distance measures are based.

Reexamining the Null Model of Microsatellite Evolution

The null model of microsatellite evolution used by the vast majority of all theoretical studies to date contains the following elements: a stepwise mutation model, an unbiased mutation process in which gains in length occur as often as losses, a constant mutation rate that does not vary with allele length or between lineages, and either a reflecting length ceiling or no restriction on allele length. There is now sufficient evidence to make at least a qualitative assessment of the validity of each of these assumptions.

The stepwise mutation model

The stepwise mutation model, in which most mutations involve the gain or loss of only a single repeat unit, is arguably the least controversial feature of microsatellite evolution. Studies that compare the theoretical expectations of this model with empirically derived allele frequency distributions find a generally good fit (Cooper et al. 1996; Shriver et al. 1993; Valdes et al. 1993). This fit can be improved by allowing occasional larger jumps (Di Rienzo *et al.* 1994). Empirical studies also support this model in that almost all germ-line mutations and most cell line mutations reported so far are of one or two repeats (Amos *et al.* 1996; Brinkmann *et al.* 1998; Primmer *et al.* 1996; Weber and Wong 1993).

Biased or unbiased mutations.

There is widespread precedent for a biased mutation process affecting short, tandemly repeated sequences, with even a recent proposal for a molecular basis (Gordenin *et al.* 1997). Mutations at both minisatellites and trinucleotide repeats associated with diseases such as Huntington's disease appear to favour expansion, with gains in length outnumbering deletions. Direct evidence from microsatellites is scant, but available data indicate that microsatellites, too, are prone to expansion (Amos *et al.* 1996; Crawford and Cuthbertson 1996; Primmer *et al.* 1996; Primmer *et al.* 1998; Weber and Wong 1993). Combining data from humans (Amos *et al.* 1996) and swallows (Primmer *et al.* 1996) yields a ratio of 47:16 gains:losses, a highly significant deviation from parity ($p = 0.00006$, one-tailed binomial probability). Other mutations thought to have occurred in immortal cell lines after the initial sampling show a similar level of bias (Weber and Wong 1993).

The concept of mutation bias is also supported by indirect evidence from several sources (for example among Y-chromosome microsatellites (Cooper *et al.* 1996), but the one I wish to focus on here concerns the relative lengths of homologous microsatellites in related species. In an unbiased model, homologous loci sampled in related lineages are no more likely to be longer in one lineage than in the other. In order to generate a significant excess of loci that are longer in one lineage relative to their homologues in another, it is necessary to have both biased mutation, to provide directionality, and variation in mutation rate, to provide a differential (see below). Thus, the finding that human microsatellites are almost invariably (33 of 40 (Rubinsztein *et al.* 1995a) and 9 of 10 (Bowcock *et al.* 1994)) longer than their homologues in chimpanzees (*Pan troglodytes*) was interpreted as evidence for both directionality and variation in rate (Rubinsztein *et al.* 1995a). Indeed, the same pattern of greater length in humans has also been observed elsewhere (Meyer *et al.* 1995) and at other tandem repeat loci such as minisatellites (Gray and Jeffreys 1991) and trinucleotide repeats (Djian *et al.* 1996; Rubinsztein *et al.* 1994) where mutation bias is known to occur.

Although genuinely consistent patterns of length difference between species require both directional evolution and variation in mutation rate, a note of caution is required. It has been suggested that such a pattern could be artefactual, arising from the preference for using long microsatellites as markers (Ellegren *et al.* 1995), an assertion which has sparked considerable debate. Support for an ascertainment bias comes from a study in which reciprocal tests were made comparing homologous loci in sheep and cattle, using markers cloned from both species (Ellegren *et al.* 1997) and showing that most markers are longer in the species from which they were cloned. However, in contrast to the human-chimpanzee comparison, which included only loci that were polymorphic in both species (Rubinsztein *et al.* 1995a), half the markers used by Ellegren *et al.* were monomorphic in one species. Since monomorphic loci are expected to be shorter under both competing hypotheses, this undermines their conclusion that a clear ascertainment bias has been demonstrated.

In a larger study based on 472 markers from the same two species, cattle and sheep, Crawford *et al.* (1998) were careful to partition their data into those loci that are polymorphic in both species and those where one homologue is monomorphic. Among markers in which both homologues are polymorphic, there is a highly significant trend for greater length in sheep, regardless of the species from which the locus was cloned. Conversely, when one homologue is monomorphic, a profound ascertainment bias operates, with most markers being longer in the species from which they were cloned. More recently, the reciprocal human-chimpanzee study has also been carried out, showing that although some ascertainment bias is present, the majority of the observed length difference is not artefactual (Cooper *et al.* 1998).

The elegant study by Crawford *et al.* (1998) thus lends twofold support to the concept of biased mutation. First, the finding that polymorphic microsatellites in sheep are longer than their homologues in cattle regardless of marker origin shows that any ascertainment bias is at best rather weak, and that, in these species at least, microsatellites are subject to directional evolution and variation in mutation rate. Second, the finding that monomorphic loci are almost invariably shorter than their polymorphic homologues is exactly what would be expected in a system prone to general expansion. Monomorphic loci have, by implication, suffered a sharp drop in mutation rate, leaving homologues in other lineages free to overtake them in length.

Variation in mutation rate and heterozygote instability

The finding of consistent length differences between species, particularly when verified by reciprocal marker testing, implies that the average mutation rate over all potential marker loci in the genome varies between lineages. This is not to be confused with locus-specific factors such as point mutations within the microsatellite repeat array, which can cause a change in the mutation rate of individual alleles (Jin *et al.* 1996). Various mechanisms have been proposed that could achieve a change in the global microsatellite mutation rate within a species, and these include changes in enzymes associated with DNA repair and DNA replication, and variation in the number of cell divisions during sperm maturation (Rubinsztein *et al.* 1995a). However, my personal preference is for the more exciting possibility that heterozygotes are more mutable than homozygotes (Amos *et al.* 1996), and this is the concept I wish to consider more fully here.

During meiosis, pairing between homologous chromosomes can lead to the "repair" of heterozygous sites through gene-conversion-like events (Borts and Haber 1989; Borts *et al.* 1990; Szostak *et al.* 1983). If heterozygous microsatellites are affected in this way, the DNA synthesis associated with a gene conversion event would provide an extra opportunity for mutation that would not affect homozygous sites to the same extent. Preliminary evidence offers support for this model, in that mutations appear more likely to arise in parents with a larger rather than smaller difference in length between the alleles they carry (Amos *et al.* 1996), and homozygous inbred lines of *Drosophila* show unexpectedly low mutation rates, up to two orders of magnitude lower than estimates for other species (Schug *et al.* 1997).

Heterozygote instability predicts a feedback loop, with any increase in heterozygosity tending to cause an increase in mutation rate, which in turn increases heterozygosity. If such a system operates, the relationship between variability and population size would be much more complex than is currently assumed. Specifically, heterozygosity would no longer correlate with population size *per se*. Instead, population size would act to modulate the rate of change of heterozygosity, with large populations experiencing a greater rate of change than equivalent smaller ones, and population expansion acting to accelerate evolution. In addition, it would no longer be correct to assume a simple relationship between genetic distance and time, since the rate of divergence would now correlate with effective population size. These properties provide predictions that allow further testing of the validity of the concept.

If evolutionary rate correlates with population size, a tendency to expand should cause large populations to carry longer alleles relative to their homologues in related, smaller populations. An obvious candidate species test-pair for prediction are humans, who have experienced dramatic population expansion, and chimpanzees, whose populations have remained relatively stable. As predicted, human minisatellites (Gray and Jeffreys 1991), trinucleotide repeat disease loci (Djian *et al.* 1996;

Rubinsztein *et al.* 1994) and microsatellites all tend to be longer than their chimpanzee homologues (Bowcock *et al.* 1994; Meyer *et al.* 1995; Rubinsztein *et al.* 1995a).

Similar trends are beginning to be documented in other species. In a study of two closely related marsupials, the northern and southern hairy-nosed wombats, 10 of 15 microsatellites had a greater mean allele length in the larger of the two populations (Taylor *et al.* 1994). Rats (Beckmann and Weber 1992) and barn swallows (Ellegren *et al.* 1995) are both abundant relative to related species and carry longer microsatellites. Sheep microsatellites are longer than their bovine homologues (Crawford *et al.* 1998), and cattle probably have an extremely small effective population size due to the dominant male contribution of "prize bulls." A last example suggests that the effect may operate among populations just as it seems to between species. A small sample of four humpback whale microsatellites were consistently longer in the (historically) larger Antarctic populations than their homologues in the smaller, north Atlantic and north Pacific populations (Valsecchi *et al.* 1997).

The heterozygote instability model also predicts increased mutation rates in hybrid zones where heterozygosity is likely to be exceptionally high, and again we find empirical support. First, many species of fur seal were hunted to the brink of extinction during the nineteenth century, and should, if anything, be genetically impoverished. However, microsatellites cloned from the grey (*Halichoerus grypus*) and harbour seals (*Phoca vitulina*) show exceptionally high levels of polymorphism in the fur seal, *Arctocephalus gazella*. Two loci are spectacular, having 6 and 7 alleles in the species from which they were cloned, yet both are longer and carry 18 and 26 alleles respectively in *A. gazella* (Gemmell *et al.* 1997), a species that is known to hybridise with *A. australis*. Second, the phenomenon of "hybrizymes" suggests that heterozygote instability in not restricted to tandem repeats. Among the many protein isozyme studies of hybrid zones, the finding of rare alleles that are not present in either parent population (Barton *et al.* 1983; Bradley *et al.* 1993) is so common that such alleles have earned their own nickname, hybrizymes (Woodruff 1989). Hybrizymes could be the products of recombination between dissimilar alleles from the two parent populations, but DNA sequencing indicates that they tend to be new mutations (Hoffman and Brown 1995; Bradley, 1993), and hence that mutation rates are elevated in hybrid zones.

Length boundaries

Long microsatellites are rare, and it is unclear why this is so. Two alternatives appear possible: a reflecting boundary, whereby expansion is prevented without eliminating the locus, for example by selection against long alleles; and an absorbing boundary, whereby above a certain length threshold loci either become unstable and delete or accumulate internal point mutations and degenerate into random sequence. Of these, only the latter seems logically compatible with general expansion, since a reflecting boundary would create a length trap at which all microsatellites would accumulate, and there is nothing yet published to suggest that such a length trap exists. Also, a reflecting upper-length boundary would cause negatively skewed allele length frequency distributions at long loci that interact with it, yet anecdotally the exact opposite seems to hold. Very long loci almost invariably show strongly positively skewed allele length frequency distributions (e.g., García de León *et al.* 1997).

Predictions of Microsatellite Allele Frequency Distributions

Regardless of the rules governing the evolution of microsatellites, the number and lengths of loci in a genome must reflect a dynamic equilibrium between the rate at which new loci are "born," the rate at which old loci are lost, and the average time spent by a locus at each length during its lifetime. As a result, the genome wide distribution of locus lengths can be thought of as an average life history, and models of microsatellite evolution should be able to account for the shape of this distribution both within and between species.

The heterozygote instability/mutation bias model makes a number of predictions about how factors such as population size will affect the distribution and lengths of microsatellites in the genome. Consider how the life of a microsatellite might run. New microsatellites are probably "born" short (Messier *et al.* 1996), from regions of high cryptic simplicity (Tautz *et al.* 1986), and then begin to expand under biased mutation. They then enter the feedback loop in which the interaction between mutation rate and heterozygosity accelerates the expansion. On reaching or exceeding a certain length, the locus becomes unstable and either deletes or disintegrates. Such a pattern would then be modified by the following species characteristics.

Effective population size

In large populations, the rate of change of heterozygosity is greater than in smaller populations, with the result that average microsatellite longevity is reduced (the life cycle completes faster and is therefore shorter). Also, since the runaway expansion will tend to begin earlier and progress faster, the average microsatellite spends less time occupying longer-length classes. Thus, for any given microsatellite birthrate, a large population will tend to carry a smaller standing crop of microsatellites, manifest as a lower density of microsatellites per kilobase of genome, and, at equilibrium, the average length of these microsatellites will be shorter than in smaller populations. It is important to stress the word "equilibrium," because rapid population expansion will cause a transient *increase* in mean length.

Body temperature

Microsatellites appear to mutate by a process of slippage, which, as with many other molecular processes, might be expected to increase in rate with temperature. Consequently, a relationship may exist between microsatellite life-history profiles and body temperature. More specifically, one could speculate that mutation rates will correlate with body temperature and, perhaps, that if the upper-length constraint is formed by a stability threshold, that this threshold might be lower in warm-blooded species. Thus, for any given population size, species with a higher body temperature might be expected to have fewer, shorter microsatellites.

The effect of sex

In otherwise equivalent species, heterozygosity levels may be reduced considerably by the mating system. For example, obligate parthenogens are completely homozygous, a factor that could slow microsatellite evolution. However, there are many less extreme examples. Most aphid species are predominantly asexual but reproduce sexually once a year, and in social insects the female line is often continued by only a tiny fraction of the true population. In such cases, heterozygosity will be reduced, predicting longer and more abundant microsatellites relative to expectations based on population size and body temperature alone.

What do Published Data Show?

Although by far the most experimental effort has been directed towards mammals, enough data have now been published on other species to test these predictions, and early indications are that the predicted and observed patterns agree. Four broad taxonomic groups can be defined based on body temperature and population size: birds, mammals, fish, and insects. Of these, the vertebrate groups can be thought of as having relatively small population sizes and high, medium, and low body temperatures, respectively, while the insects have generally large population sizes and low body temperatures.

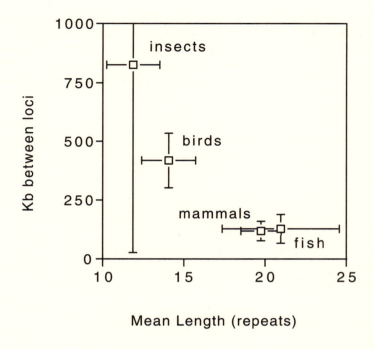

Figure 4.1. Scatter plot of the distribution of microsatellite lengths and genomic frequency for four groups of organisms. These are insects ($n = 31$ loci), birds ($n = 33$), mammals ($n = 112$) and cold-blooded vertebrates ($n = 28$). Data are based primarily on all papers published in the journal *Molecular Ecology* prior to July 1997, but augmented by DNA sequences deposited in the EMBL database and Genbank. Only dinucleotide repeat motifs are included. Kilobases per locus were calculated from data on the proportion of positively hybridizing recombinant clones and the mean size of insert DNA used in library construction (only given in a minority of studies). Errors bars are standard errors of the mean.

To investigate the distribution and lengths of microsatellites in diverse species, I conducted an extensive but by no means exhaustive literature search, concentrating on the journal *Molecular Ecology*, but also interrogating the European Molecular Biology Laboratory (EMBL) database and citation indexes. The results are summarized in Figure 4.1. When the length and frequency of microsatellites in these groups are examined it is found that, among vertebrates, both mean length and frequency per kilobase decline with increasing body temperature. Indeed, the rarity and brevity of bird dinucleotide repeats have already been noted (Primmer *et al.* 1997), and the two most impressive microsatellite longevity records are both held by cold-blooded vertebrates (FitzSimmons *et al.* 1995; Rico *et al.* 1996). Interestingly, whales have relatively low body temperatures compared with other mammals, and 4 of 20 whale microsatellites isolated in my laboratory carry longer alleles than can be found among more than 400 human markers (pers. obs.). Insects have rather few microsatellites, most of which are short. It is also intriguing that the majority of successful insect studies involve species which are likely to have unusually low heterozygosities due to their breeding system, being either eusocial or cyclically parthenogenetic (ants, Ross *et al.* 1997; wasps, Strassmann *et al.* 1998; aphids, Sunnocks *et al.* 1996).

Of course, published studies are heavily biased in favour of successful isolation and variable loci. There is also the problem that most people stop isolating new loci when they have sufficient ones, rather than expending a constant experimental effort, and that people vary in how proficient they are. Less skilled scientists working on species in which microsatellites are abundant may report similar frequencies to those reported by experimentally gifted researchers working with species in which microsatellites are rare. Failure to find dinucleotide repeats may cause researchers to throw their net a little wider and look for tri- and tetranucleotide motifs. Thus, in my literature search I found that only one of 95 mammalian markers did not have a dinucleotide motif, while over 75% of all insect and bird markers were trinucleotides. All these factors can confound and confuse, in general acting to reduce the differences between studies. That a trend exists at all, therefore, suggests to me that the underlying pattern is strong.

4.3 CONCLUSIONS

There can be no doubt that genetic analysis can play an extremely positive role in conservation. It can help to identify species. It can help define population boundaries, reveal patterns of migration, and elucidate various forms of social organization. Through parentage testing it can uncover mating behaviours. All these pieces of knowledge can aid in the definition and understanding of what we are trying to conserve. However, it is vitally important that we do not go too far in accepting without question the potential benefits, particularly in areas where other aspects of biology are poorly understood. A prime case in question is the relationship between genetic variability and evolutionary health, where much is made of species with low variability that appear to be struggling, while species such as the badger and the northern elephant seal, with low variability but healthy, dramatically recovering populations are too frequently ignored.

The two sections of this article thus present two different ways in which there may be problems in the interpretation of genetic data. On the one hand, it seems that there is a general tendency to accredit too much power to genetic drift during bottlenecks in removing variability, and on the other, we clearly need to learn much more about microsatellite evolution in order to interpret correctly the vast amounts of data that have been, are being, and will be collected. As such, the take-home message is undeniably negative. However, there are positive elements as well.

By recognizing the problems inherent in trying to assess what are normal levels of variability and how much depletion can be attributed to bottlenecks, we will be in a much better position to get at the truth. Inclusion of more indications of experimental effort (number of markers tested and failed attempts) in publications would help, as would the use of ancient DNA techniques to access samples prior to population decline. Greater use of direct calculation or computer simulations to assess plausible levels of depletion should also be encouraged, since these would expose the levels of loss that have to be attributed to other factors, such as natural selection.

One of the most exciting recent revelations is a new method for quantifying inbreeding, termed mean d^2 (Coulson *et al.* 1998). The method takes advantage of the fact that independently evolving microsatellite alleles tend to diverge in length with time. Briefly, at any one locus, the length difference between the two alleles carried by an individual provides a measure of relatedness between the two parental chromosomes on which that locus sits. Averaged over many loci, such length differences provide a measure of relatedness between the two parental genomes, or inbreeding. Instead of quantifying inbreeding simply in terms of the proportion of loci that are heterozygous, this new method provides a continuous scale on which heterozygosity can be measured. Studies on harbour seals and red deer show that the increased resolution afforded can reveal fitness components associated with inbreeding that do not show up with more traditional analyses (Coltman *et al.* 1998; Coulson *et al.* 1998). Such studies promise to revolutionize our ability to study the link between population decline and fitness.

The concepts raised during discussion of how microsatellites evolve appear worrying in that, if correct, they would necessitate a rethinking of how to assess variability and genetic distance. Yet, the problems are not so very profound. Populations that have become isolated will still tend to accumulate differences, while those that mix freely will appear similar. It is only the measurement of distance in populations that differ greatly in size that may suffer. At the same time, the possibility of heterozygote instability would create exciting new tools with which to study evolutionary processes. For example, historical population-size differences could be inferred from a disparity between the haploid mitochondrial and diploid nuclear marker divergence or diversity. The human-chimpanzee comparison provides a possible case in question, in that chimpanzees have greater mitochondrial diversity but lower nuclear diversity (Wise *et al.* 1997), as would be expected if the recent human expansion had caused a disproportionate increase in nuclear diversity. Similarly, instead of inferring heterozygote instability from increased microsatellite length and diversity in known hybrids, it might become possible to turn the tables and use microsatellite data to infer where hybridisation was occurring. In the meantime, there is an urgent need for new empirical studies aimed at confirming or indicating a change in the patterns that seem to exist.

REFERENCES

Amos, W., Barrett, J. A., and Dover, G. A. (1991). Breeding behaviour of pilot whales revealed by DNA fingerprinting. *Heredity*, *67*, 49–55.

Amos, W., and Harwood, J. (1998). Factors affecting levels of genetic diversity in natural populations. *Phil. Trans. R. Soc. Lond. B*, *353*, 177–186.

Amos, W., and Pemberton, J. M. (1992). DNA fingerprinting in non-human populations. *Curr. Opin. Gen. Develop.*, *2*, 857–860.

Amos, W., Sawcer, S. J., Feakes, R., and Rubinsztein, D. C. (1996). Microsatellites show mutational bias and heterozygote instability. *Nature Genet.*, *13*, 390–391.

Amos, W., Schlötterer, C., and Tautz, D. (1993). Social structure of pilot whales revealed by analytical DNA typing. *Science*, *260*, 670–672.

Avise, J. C., Bowen, B. W., and Lamb, T. (1989). DNA fingerprints from hypervariable mitochondrial genotypes. *Mol. Biol. Evol.*, *6*, 258–269.

Baker, C. S., Palumbi, S. R., Lambertsen, R. H., Weinrich, M. T., Calambokidis, J., and O'Brien, S. J. (1990). Influence of seasonal migration on geographic distribution of mitochondrial DNA haplotypes in humpback whales. *Nature*, *344*, 238–240.

Barrett, S. C. H., and Charlesworth, D. (1991). Effects of a change in the level of inbreeding on the genetic load. *Nature*, *352*, 522–524.

Barton, N. H., Halliday, R. B., and Hewitt, G. M. (1983). Rare electrophoretic variants in a hybrid zone. *Heredity*, *50*, 139–146.

Beckmann, J. S., and Weber, J. L. (1992). Survey of rat and human microsatellites. *Genomics*, *12*, 627–631.

Bonner, W. N. (1968). The fur seal of South Georgia. *Brit. Ant. Survey Sci. Rep.*, *56*, 1–81.

Borts, R. H., and Haber, J. E. (1989). Length and distribution of meiotic gene conversion tracts and crossovers in *Saccharomyces cerevisiae*. *Genetics*, *123*, 69–80.

Borts, R. H., Leung, W.-Y., Kramer, W., Kramer, B., Williamson, M., Fogel, S., and Haber, J. E. (1990). Mismatch repair-induced meiotic recombination requires the *PMS1* gene product. *Genetics*, *124*, 573–584.

Bowcock, A. M., Ruiz Linares, A., Tomfohrde, J., Minch, E., Kidd, J. R., and Cavalli-Sforza, L. L. (1994). High resolution trees with polymorphic microsatellites. *Nature, Lond, 368*, 455–457.

Bradley, R. D., Bull, J. J., Johnson, A. D., and Hillis, D. M. (1993). Origin of a novel allele in a mammalian hybrid zone. *Proc. Natl. Acad. Sci. USA, 90*, 8939–8941.

Brinkmann, B., Klintschar, M., Neuhuber, F., Hühne, J., and Rolf, B. (1998). Mutation rate in human microsatellites: Influence of the structure and length of the tandem repeat. *Am. J. Hum. Genet., 62*, 1408–1415.

Brooker, M. G., Rowley, I., Adams, M., and Baverstock, P. R. (1990). Promiscuity: An inbreeding avoidance mechanism in a socially monogamous species. *Behav. Ecol. Sociobiol., 26*, 191–199.

Brookes, M. I., Graneau, Y. A., King, P., Rose, O. C., Thomas, C. D., and Mallet, J. L. B. (1997). Genetic analysis of founder bottlenecks in the rare British butterfly *Plebejus argus*. *Conserv. Biol., 11*, 648–661.

Bruford, M. W., and Wayne, R. K. (1993). Microsatellites and their application to population genetic studies. *Curr. Opin. Gen. Develop., 3*, 939–943.

Burke, T. (1989). DNA fingerprinting and other methods for the study of mating success. *Trends Ecol. Evol., 4*, 139–144.

Caro, T. M. (1994). *Cheetahs of the Serengeti plains*. Chicago: University of Chicago Press.

Caro, T. M., and Laurenson, M. K. (1994). Ecological and genetic factors in conservation: A cautionary tale. *Science, 263*, 485–486.

Coltman, D. W., Bowen, W. D., and Wright, J. M. (1998). Birth weight and neonatal survival of harbour seal pups are positively correlated with genetic variation measured by microsatellites. *Proc. R. Soc. Lond. B, 265*, 803–809.

Cooper, G., Amos, W., Hoffman, D., and Rubisztein, D. C. (1996). Network analysis of human Y microsatellite haplotypes. *Hum. Mol. Genet., 5*, 1759–1766.

Cooper, G., Rubinsztein, D. C., and Amos, W. (1998). Ascertainment bias does not entirely account for human microsatellites being longer than their chimpanzee homologues. *Hum. Mol. Genet., 7(9)*, 1425–1429.

Coulson, T. N., Pemberton, J. M., Albon, S. D., Beaumont, M., Marshall, T. C., Slate, J., Guiness, F. E., and Clutton-Brock, T. H. (1998). Microsatellites reveal heterosis in red deer. *Proc. R. Soc. Lond. B, 265*, 489–495.

Crawford, A., Knappes, S. M., Paterson, K. A., deGotari, M. J., Dodds, K. G., Freking, R. T., Stone, R. T., and Beattie, C. W. (1998). Microsatellite evolution: Testing the ascertainment bias hypothesis. *J. Mol. Evol., 46*, 256–260.

Crawford, A. M., and Cuthbertson, R. P. (1996). Mutations in sheep microsatellites. *Genome Res., 6*, 876–879.

Deka, R., Shriver, M. D., Yu, L. M., Aston, C. E., Chakraborty, R., and Ferrell, R. (1994). Conservation of human chromosome 13 polymorphic microsatellite $(CA)_n$ repeats in chimpanzees. *Genomics, 22*, 226–230.

Di Rienzo, A., Peterson, A. C., Garza, J. C., Valdes, A. M., and Slatkin, M. (1994). Mutational processes of simple sequence repeat loci in human populations. *Proc. Natl. Acad. Sci. USA, 91*, 3166–3170.

Djian, P., Hancock, J. M., and Chana, H. S. (1996). Codon repeats in genes associated with human diseases: Fewer repeats in the genes of nonhuman primates and nucleotide substitutions concentrated at sites of reiteration. *Proc. Natl. Acad. Sci. USA, 93*, 417–421.

Dow, B. D., and Ashley, M. V. (1996). Microsatellite analysis of seed dispersal and parentage of saplings in bur oak, *Quercus macrocarpa*. *Mol. Ecol., 5*, 615–627.

Ellegren, H., Moore, S., Robinson, N., Byrne, K., Ward, W., and Sheldon, B. C. (1997). Microsatellite evolution — a reciprocal study of repeat lengths at homologous loci in cattle and sheep. *Mol. Biol. Evol.*, *14*, 854–860.

Ellegren, H., Primmer, C. R., and Sheldon, B. C. (1995). Microsatellite evolution: Directionality or bias in locus selection. *Nature Genet.*, *11*, 360–362.

FitzSimmons, N. N., Moritz, C., and Moore, S. S. (1995). Conservation and dynamics of microsatellite loci over 300 million years of marine turtle evolution. *Mol. Biol. Evol.*, *12*, 432–440.

García de León, F. J., Chikhi, L., and Bonhomme, F. (1997). Microsatellite polymorphism and population subdivision in natural populations of European sea bass *Dicentrarchus labrax* (Linnaeus, 1758). *Mol. Ecol.*, *6*, 51–62.

Garza, J. C., Slatkin, M., and Freimer, N. B. (1995). Microsatellite allele frequencies in humans and chimpanzees with implications for constraints on allele size. *Mol. Biol. Evol.*, *12*, 594–604.

Gemmell, N. J., Allen, P. J., Goodman, S. J., and Reed, J. Z. (1997). Interspecific microsatellite markers for the study of pinniped populations. *Mol. Ecol.*, *6*, 661–666.

Gilbert, D. A., Lehman, N., O'Brien, S. J., and Wayne, R. K. (1990). Genetic fingerprinting reflects population differentiation in the California Channel Island fox. *Nature*, *344*, 764–767.

Gilbert, D. A., Packer, C., Pusey, A. E., Stephens, J. C., and O'Brien, S. J. (1991). Analytical DNA fingerprinting in lions: Parentage, genetic diversity and kinship. *J. Hered.*, *82*, 378–386.

Gilpin, M. (1991). The genetic effective size of a metapopulation. *Biol. J. Linn. Soc.*, *42*, 165–175.

Goldstein, D. B., and Pollock, D. D. (1997). Launching microsatellites: A review of mutation processes and methods of phylogenetic inference. *J. Hered.*, *88*, 335–342.

Goldstein, D. B., Ruiz Linares, A., Cavalli-Sforza, L. L., and Feldman, M. W. (1995a). An evaluation of genetic distances for use with microsatellite loci. *Genetics*, *139*, 463–471.

Goldstein, D. B., Ruiz Linares, A., Cavalli-Sforza, L. L., and Feldman, M. W. (1995b). Genetic absolute dating based on microsatellites and the origin of modern humans. *Proc. Natl. Acad. Sci. USA*, *92*, 6723–6727.

Gordenin, D. A., Kunkel, T. A., and Resnick, M. A. (1997). Repeat expansion — all in a flap. *Nature Genet.*, *16*, 116–118.

Gray, I. C., and Jeffreys, A. J. (1991). Evolutionary transience of hypervariable minisatellites in man and the primates. *Proc. R. Soc. Lond. B*, *243*, 241–253.

Hagelberg, E. (1994). Ancient DNA studies. *Evol. Anthropol.* 2 (6), 199–207.

Hagelberg, E., Thomas, M. G., Cook Jr, C. E., Sher, A. V., Baryshnikov, G. F., and Lister, A. M. (1994). DNA from ancient mammoth bones. *Nature*, *370*, 333–334.

Hartl, D. L. (1988). *A primer of population genetics*. Sunderland, Mass.: Sinauer.

Hedrick, P. W. (1995). Elephant seals and the estimation of a population bottleneck. *J. Hered.*, *86*, 232–235.

Hedrick, P. W. (1996). Bottleneck(s) or metapopulation in cheetahs. *Conserv. Biol.*, *10*, 897–899.

Higuchi, R., Bowman, B., Freiberger, M., Ryder, O. A., and Wilson, A. C. (1984). DNA sequences from the quagga, an extinct member of the horse family. *Nature*, *312*, 287–289.

Hillis, D. M., and Moritz, C. (1990). An overview of applications of molecular systematics. In Hillis, D. M., and Moritz, C. (eds.), *Molecular systematics*, 508–515. Sunderland, Mass.: Sinauer.

Hoelzel, A. R., Halley, J., and O'Brien, S. J. (1993). Elephant seal genetic variation and the use of simulation models to investigate historical population bottlenecks. *J. Hered.*, *84*, 443–449.

Hoffman, S. M. G., and Brown, W. M. (1995). The molecular mechanism underlying the rare allele phenomenon in a subspecific hybrid zone of the California field-mouse, *Peromyscus californicus*. *J. Mol. Evol.*, *41*, 1165–1169.

Jeffreys, A. J., Allen, M. J., Hagelberg, E., and Sonnberg, A. (1992). Identification of the skeletal remains of Josef Mengele by DNA analysis. *For. Sci. Int.*, *56*, 65–76.

Jeffreys, A. J., Wilson, V., and Thein, S. L. (1985a). Hypervariable "minisatellite" regions in human DNA. *Nature, Lond.*, *314*, 67–73.

Jeffreys, A. J., Wilson, V., and Thein, S. L. (1985b). Individual-specific "fingerprints" of human DNA. *Nature, Lond.*, *316*, 76–79.

Jin, L., Macaubas, C., Hallmayer, J., Kimura, A., and Mignot, E. (1996). Mutation rate varies among alleles at a microsatellite locus: Phylogenetic evidence. *Proc. Natl. Acad. Sci. USA*, *93*, 15285–15288.

Keller, L. F., Arcese, P., Smith, J. N. M., Hochachka, W. M., and Stears, S. C. (1994). Selection against inbred song sparrows during a natural population bottleneck. *Nature*, *372*, 356–357.

Kipling, D., and Cooke, H. J. (1990). Hypervariable ultra-long telomeres in mice. *Nature*, *347*, 400–402.

Lande, R., and Barrowclough, G. F. (1987). Effective population size, genetic variation, and their use in population management. In Soulé, M. E. (ed.), *Viable populations for conservation*, 189, Cambridge: Cambridge University Press.

Laurenson, M. K., Caro, T. M., Gros, P., and Wielebnowski, N. (1995). Controversial cheetahs? *Nature*, *377*, 392.

Leberg, P. L. (1992). Effects of population bottlenecks on genetic diversity as measured by allozyme electrophoresis. *Evolution*, *46*, 447–494.

Macdonald, D. (1984). *The encyclopaedia of mammals*. London: Allen and Unwin.

Majerus, M. E. N., Amos, W., and Hurst, G. D. D. (1996). *Evolution, the four billion year war*. Harlow, Essex: Longman.

Messier, W., Li, S.-H., and Stewart, C.-B. (1996). The birth of microsatellites. *Nature*, *381*, 483.

Meyer, E., Wiegand, P., Rand, S. P., Kuhlmann, D., Brack, M., and Brinkmann, B. (1995). Microsatellite polymorphisms reveal phylogenetic relationships in primates. *J. Mol. Evol.*, *41*, 10–14.

Moritz, C. (1994). Applications of mitochondrial DNA analysis in conservation: A critical review. *Mol. Ecol.*, *3*, 401–411.

Mundy, N. I., Winchell, C. S., Burr, T., and Woodruff, D. S. (1997). Microsatellite variation and microevolution in the critically endangered San Clemente Island loggerhead shrike (*Lanius ludovicianus mearnsi*). *Proc. R. Soc. Lond. B*, *264*, 869–875.

Nauta, M. J., and Weissing, F. J. (1996). Constraints on allele size at microsatellite loci: Implications for genetic differentiation. *Genetics*, *143*, 1021–32.

Nevo, E., Bieles, A., and Schlomo, B. (1984). The evolutionary significance of genetic diversity: Ecological, demographic and life history consequences. In G. S. Mani (ed.), *Evolutionary dynamics of genetic diversity*, Berlin: Springer-Verlag.

O'Brien, S. J. (1994). A role for molecular genetics in biological conservation. *Proc. Natl. Acad. Sci. USA*, *91*, 5748–55.

Packer, C. A., Gilbert, D. A., Pusey, A. E., and O'Brien, S. J. (1991). A molecular genetic analysis of kinship and cooperation in African lions. *Nature*, *351*, 562–65.

Paetkau, D., Calvert, W., Stirling, I., and Strobeck, C. (1995). Microsatellite analysis of population structure in Canadian polar bears. *Mol. Ecol.*, *4*, 347–54.

Primmer, C., Ellegren, H., Saino, N., and Møller, A. P. (1996). Directional evolution in germline microsatellite mutations. *Nature Genet.*, *13*, 391–93.

Primmer, C. R., Raudsepp, T., Chowdhary, B. P., Møller, A. P., and Ellegren, H. (1997). Low frequency of microsatellites in the avian genome. *Genome Res.*, *7*, 471–82.

Primmer, C. R., Saino, N., Møller, A. P., and Ellegren, G. (1998). Unravelling the process of microsatellite evolution through analysis of germ line mutations in barn swallows *Hirundo rustica*. *Mol. Biol. Evol. 15*, 1047–54.

Queller, D. C., Strassmann, J. E., and Hughes, C. R. (1993). Microsatellites and kinship. *Trends Ecol. Evol.*, *8*, 285–88.

Ralls, K., Brugger, K., and Ballou, J. (1970). Inbreeding and juvenile mortality in small populations of ungulates. *Science*, *206*, 1101–3.

Rico, C., Rico, I., and Hewitt, G. (1996). 470 million years of conservation of microsatellite loci among fish species. *Proc. R. Soc. Lond. B*, *263*, 549–57.

Ross, K. G., Krieger, M. J. B., Shoemaker, D. D., Vargo, E. L., and Keller, L. (1997). Hierarchical analysis of genetic structure in native fire ant populations: Results from three classes of molecular markers. *Genetics*, *151*, 545–63.

Rubinsztein, D. C., Amos, W., Leggo, J., Goodburn, S., Jain, S., Li, S. H., Margolis, R. L., Ross, C. A., and Ferguson-Smith, M. (1995a). Microsatellites are generally longer in humans compared to their homologues in non-human primates: Evidence for directional evolution at microsatellite loci. *Nature Genet.*, *10*, 337–43.

Rubinsztein, D. C., Amos, W., Leggo, J., Goodburn, S., Ramesar, R. S., Old, J., Bontrop, R., McMahon, R., Barton, D. E., and Ferguson-Smith, M. A. (1994). Mutational bias provides a model for the evolution of Huntington's disease and predicts a general increase in disease prevalence. *Nature Genet. 7*, 525–30.

Rubinsztein, D. C., Leggo, J., and Amos, W. (1995b). Microsatellites evolve more rapidly in humans than in chimpanzees. *Genomics*, *30*, 610–612.

Saccheri, I. J., Brakefield, P. M., and Nichols, R. A. (1996). Severe inbreeding depression and rapid fitness rebound in the butterfly *Bicyclus anynana* (Satyridae). *Evolution*, *50*, 2000–2013.

Saccheri, I. J., Kuussaari, M., Vikman, P., Fortelius, W., and Hanski, I. (1998). Inbreeding and extinction in a butterfly metapopulation. *Nature*, *392*, 491–494.

Schlötterer, C., and Tautz, D. (1992). Slippage synthesis of simple sequence DNA. *Nucleic Acids Res.*, *20*, 211–215.

Schug, M. D., Mackay, T. F. C., and Aquadro, C. F. (1997). Low mutation rates of microsatellites in *Drosophila melanogaster*. *Nature Genet.*, *15*, 99–102.

Shriver, M. D., Jin, L., Chakraborty, R., and Boerwinkle, E. (1993). VNTR allele frequency distributions under the stepwise mutation model: A computer simulation approach. *Genetics*, *134*, 983–993.

Slade, R. W., Moritz, C., Hoelzel, A. R., and Burton, H. R. (1998). Molecular population genetics of the southern elephant seal *Mirounga leonina*. *Genetics*, *149*, 1945–1957.

Slatkin, M. (1995). A measure of population subdivision based on microsatellite allele frequencies. *Genetics*, *139*, 457–462.

Stewart, B. S., Yochem, P. K., Huber, H. R., DeLong, R. L., Jameson, R. J., Sydeman, W. J., Allen, S. G., and Le Boeuf, B. J. (1994). History and present status of the northern elephant seal population. In Le Boeuf, B. J., and Laws, R. M. (eds.), *Elephant seals: Population ecology, behaviour and physiology*, Berkeley, California: University of California Press.

Strassmann, J. E., Goodnight, K. F., Klingler, C. J., and Queller, D. C. (1998). The genetic structure of swarms and the timing of their production in the queen cycles of neotropical wasps. *Mol. Ecol.*, *7*, 709–718.

Sunnocks, P., England, P. R., Taylor, A. C., and Hales, D. F. (1996). Microsatellite and chromosome evolution of parthenogenetic Sitobion aphids in Australia. *Genetics*, *144*, 747–756.

Szostak, J. W., Orr-Weaver, T. L., and Rothstein, R. J. (1983). The double strand break repair model for recombination. *Cell*, *33*, 25–35.

Takezaki, N., and Nei, M. (1996). Genetic distances and reconstruction of phylogenetic trees from microsatellite data. *Genetics*, *144*, 389–399.

Tautz, D. (1989). Hypervariability of simple sequences as a general source of polymorphic DNA markers. *Nucleic Acids Res.*, *17*, 6462–6471.

Tautz, D., Trick, M., and Dover, G. A. (1986). Cryptic simplicity as a major source of genetic variation. *Nature, Lond.*, *322*, 652–656.

Taylor, A. C., Sherwin, W. B., and Wayne, R. K. (1994). Genetic variation of microsatellite loci in a bottlenecked species: The hairy-nosed wombat *Lasiorhinus krefftii*. *Mol. Ecol.*, *3*, 277–290.

Townsend, C. H. 1885 An account of recent captures of the california sea-elephant, and statistics relating to the present abundance of the species. *Proc. U.S. Nat. Mus.*, *8*, 90–93.

Valdes, A. M., Slatkin, M., and Freimer, N. B. (1993). Allele frequencies at microsatellite loci: The stepwise mutation model revisited. *Genetics*, *133*, 737–749.

Valsecchi, E., Palsbøll, P., Hale, P., Glockner-Ferrari, D., Ferrari, M., Clapham, P., Larsen, F., Mattila, D., Sears, R., Sigurjonsson, J., Brown, M., Corkeron, P., and Amos, W. (1997). Microsatellite genetic distances between oceanic populations of the humpback whale (*Megaptera novaeangliae*). *Mol. Biol. Evol.*, *14*, 355–362.

Wahls, W. P., Wallace, L. J., and Moore, P. D. (1990). Hypervariable minisatellite DNA is a hotspot for homologous recombination in human cells. *Cell*, *60*, 95–103.

Wayne, R. K., and Jenks, S. M. (1991). Mitochondrial DNA analysis implying extensive hybridisation of the endangered red wolf *Canis rufus*. *Nature*, *351*, 565–568.

Weber, J. L., and Wong, C. (1993). Mutation of human short tandem repeats. *Hum. Mol. Genet.*, *2*, 1123–1128.

Wenink, P. W., Baker, A. J., and Tilanus, M. G. J. (1993). Hypervariable control region sequences reveal global population structuring in a long-distance migrant shorebird, the dunlin (*Calidris alpina*). *Proc. Natl. Acad. Sci. USA*, *90*, 94–98.

Wielebnowski, N. (1996). Reassessing the relationship between juvenile mortality and genetic monomorphism in captive cheetahs. *Zoo Biol.*, *15*, 353–369.

Wilson, A. C., Cann, R. L., Carr, S. M., George, M., Gyllensten, U. B., Helm-Bychowski, K. M., Higuchi, R. G., Palumbi, S. R., Prager, E. M., Sage, R. D., and Stoneking, M. (1985). Mitochondrial DNA and two perspectives on evolutionary genetics. *Biol. J. Linn. Soc.*, *26*, 375–400.

Wise, C. A., Sraml, M., Rubinsztein, D. C., and Easteal, S. (1997). Comparative nuclear and mitochondrial diversity in humans and chimpanzees. *Mol. Biol. Evol.*, *14*, 707–716.

Woodruff, R. C. (1989). Genetic anomalies associated with *Cerion* hybrid zones: The origin and maintenance of new electrophoretic variants called hybrizymes. *Biol. J. Linn. Soc.*, *36*, 281–294.

5

One Use of Phylogenies for Conservation Biologists:
Inferring Population History from Gene Sequences

PAUL H. HARVEY AND HELEN STEERS

SUMMARY. We describe a route for inferring population dynamic history directly from nonrecombining gene sequence data collected from contemporary individuals. As gene sequence data become easier to collect and more widely available, we should expect the family of techniques we describe to constitute a set of useful tools for preliminary data exploration by biologists who need to know the population dynamic status of a previously undescribed population. However, there is still a need to construct new techniques to deal with some unrealistic assumptions; we describe current limitations and some possible solutions.

INTRODUCTION

Conservation biologists face an awesome task in the coming decades. The purpose of this contribution is to describe a relatively new way in which genetic and, in particular, phylogenetic information may be used to provide a useful tool that can be used as an aid to decision making.

There are several different reasons why phylogenetic information can help conservation biologists. Biodiversity can be defined in terms of genetic diversity, in which case phylogenies provide a description of biodiversity because they describe the way in which genetic diversity is allocated among taxa. If an aim of conservation biology is to maximise conserved biodiversity, it is possible to provide an objective description of the extent to which alternative management decisions that conserve some species at the expense of others do, in fact, conserve different amounts of biodiversity, not just in terms of number of species but as quantities of diverse genetic material. After a promising start in which the problem was reasonably well posed (Vane-Wright *et al.* 1991), a somewhat confused series of ad hoc measures of diversity based on phylogenetic relatedness were proposed (listed in Harvey 1995). Fortunately,

the confusion has now subsided with some general acceptance that, if
(i) biodiversity is defined in terms of genetic diversity and (ii) branch
lengths on a phylogenetic tree measure the amount of genetic change between nodes or between nodes and tips, then the conservation of those
extant species which maximise the conserved branch length on the phylogenetic tree will maximise biodiversity (although the results of Nee and
May (1997) suggest that benefits in terms of increased genetic variance
retained as a result of such a maximization are likely to be small). Of
course, conserving all species is the ideal, but if a choice of species must
be made then one criterion which might help guide that choice has been
specified. This topic has been considered at length in a series of papers,
reviews and book chapters (e.g. Williams *et al.* 1994) and we shall leave
it here. A more general review of the role of phylogenies in conservation
biology is also available elsewhere (Moritz 1995).

We shall, instead, focus on the vexed question of population history.
Conservation biologists frequently encounter contemporary populations
whose dynamic history is unknown. Yet if two populations of the same
size have different histories, such that one has been steadily growing
in numbers while the other has been plummeting, then we should do
well to concentrate our conservation efforts on the latter. If we had a
family tree showing when all pairs of individuals in a contemporary population last shared a common ancestor, it would be possible to analyse
the structure of that tree in an attempt to determine the likely population history. Using a variety of newly developed procedures, we could
determine whether there are lineages that have been statistically more
prolific than others, whether there is evidence of a generalised increase
or decrease in the net birth rate through time, or whether population
explosions or crashes have occurred. If there is no evidence for changes
in population birth and death rates, it is even possible to estimate those
birth and death rates using maximum likelihood procedures. Clearly,
in practice we are not likely to have the required genealogical tree for
a whole population, but for the moment we shall imagine that molecular genetic protocols have become so advanced that we can indeed infer
such trees. Subsequently, we shall see what population dynamic message is retained when the data become more realistically fragmentary
and messy. Finally, we shall review the assumptions behind the new
methods, the extent to which they are likely to hold in practice, and
problems on which progress has yet to be made.

A family tree reconstructed from the relationships among contemporary individuals shows only a subset of lineages that existed in the
past — those individuals that left no descendants to the present day are
not represented on the genealogy. Visually, one particularly useful technique for population analysis is to produce a lineages-through-time plot

that shows the reconstructed population history, from which we infer information about the actual population history.

5.1 When the Whole Population Has Been Sampled

Genetic material from different individuals in a population can be used to construct an intrapopulation phylogeny, or gene genealogy, which shows when pairs of individuals last shared a common ancestor. The reconstructed temporal distribution of nodes, when displayed as cumulative lineages-through-time plots can provide useful information about the dynamic history of the population being studied. In Figure 1, we provide output from a computer application called BI-DE (Rambaut *et al.* 1996), which simulates population growth under user-specified scenarios and then recreates that part of the genealogy that shows the relationships among extant individuals in the population. The number of lineages is plotted against time for the actual population as it grew, and an equivalent plot is given for the reconstructed genealogy. In most cases, except at the present, the reconstructed line will lie below the actual line, the difference between the two lines being the number of individuals that lived at some specified point in the past but did not give rise to contemporary descendants.

As is evident from Figure 5.1, different population histories give different reconstructed genealogies. When the number-of-lineages axis is scaled logarithmically, the actual and reconstructed plots coincide as a straight line under a pure birth process. However, when some mortality is introduced into the simulation, the reconstructed lines steepen towards the present. Density-dependent population growth, varied according to changing birth or death rates, produces a plot that is steep in the distant past, shallows off, and then steepens towards the present; the steep portion in the distant past distinguishes density-dependent population growth from one in which birth and death rates have remained constant through time. Finally, we give an example where birth and death rates have remained constant in time except at one point when a large proportion of the population has died; the point where a population crash occurs appears as a sharp change in slope on the reconstructed plot. The examples in Figure 5.1 are provided simply to demonstrate that reconstructed lineages-through-time plots have the potential for distinguishing among scenarios of population dynamic history. Analytical treatments are given in Harvey *et al.* (1994).

The constant-rates birth-death process, which gives rise to an exponentially growing population when the birth rate is greater than the death rate, has been the subject of fruitful analytical study. Figure 5.2

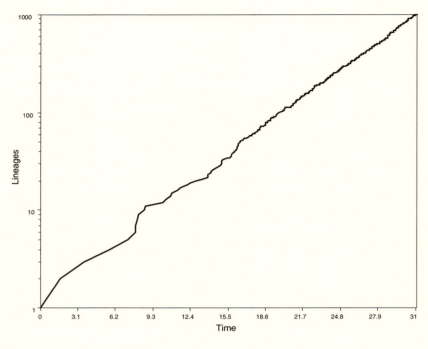

Figure 5.1a

Figure 5.1. Lineages-through-time plots for populations with different dynamic histories. All populations grow until they reach 1000. Two lines are shown on each plot: the upper line is the actual number of individuals in the population and the lower line is the apparent number reconstructed from the genealogy — the difference between the actual and apparent results from the fact that some individuals that lived in the past left no descendants in the present (with a pure birth process, the two lines are superimposed because no individuals have died). Different population histories leave characteristically shaped lineages-through-time plots (see text). The plots show simulations for (a) pure birth process: birth rate = 0.2, death rate = 0.0; (b) birth-death process: birth rate = 0.2, death rate = 0.15; (c) density dependant mortality: birth rate = 0.2, death rate = 0.1+(number of lineages)/10000; (d) density dependent fecundity: birth rate = 0.2-(number of lineages)/10000, death rate = 0.1; (e) population bottleneck: birth rate = 0.2, death rate = 0.1, population grows to 1000 individuals, randomly selected 95% of individuals die, population grows to 1000.

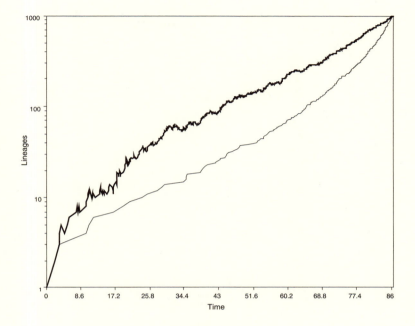

Figure 5.1b

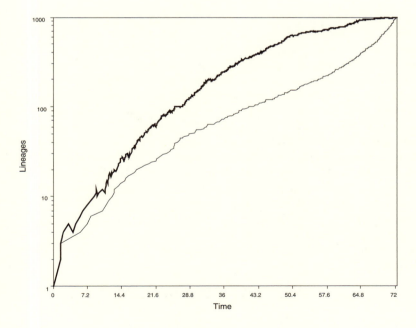

Figure 5.1c

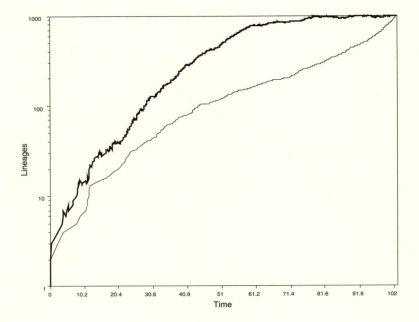

Figure 5.1d

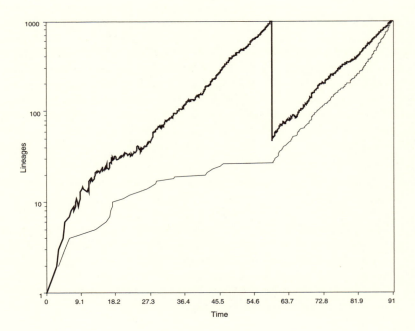

Figure 5.1e

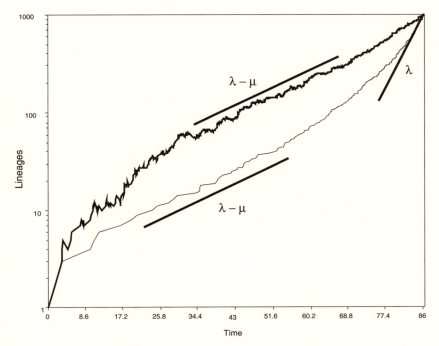

Figure 5.2. As in Figure 5.1b, a population has been growing exponentially with a birth rate λ of 0.2 and a death rate μ of 0.1. Analytical work (Harvey *et al.* 1994) reveals that parts of the reconstructed plot have slopes that, as shown, can be used to estimate population birth and death rates. The reason that the actual plot is particularly steep at the start is that only a sample of those populations that are initiated will actually grow to a reasonable size before becoming extinct; the subset that does make it to 1000 lineages contains a disproportionately high representation of those populations that, by chance, got off to a good start. The reconstructed plot steepens towards the present because individuals born in the very recent past will have had little chance to die. Maximum likelihood procedures have been developed for estimating both λ and μ.

reproduces Figure 5.1b, but adds the expected values for the slopes of both the actual and reconstructed lineages-through-time plots. Of interest here is the fact that, for reasons described in the legend to Figure 5.1, the reconstructed plot steepens towards the present. In fact, it can be shown analytically (Harvey *et al.* 1994) that the slope over the linear portion of the graph is approximately the net birth rate (birth rate mi-

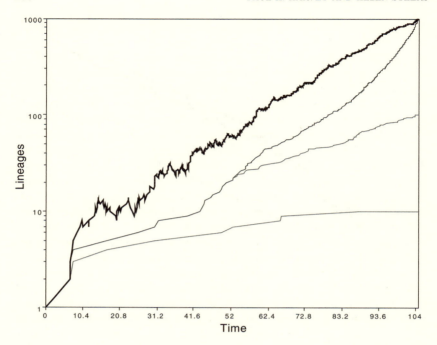

Figure 5.3. When not all members of a local population are compared in the construction of a genealogy, the more recent nodes tend to be relatively undersampled because descendants from both daughter branches are less likely to be represented in a sample than those for more ancient nodes. Here, a simulation has been performed with a birth rate of 0.2 and a death rate of 0.15, and the population has grown to 1000 as in Figure 5.1b. The actual phylogeny is represented by the top line, the reconstructed lineages-through-time plot for the full sample of contemporary individuals is given by the second line down. When only 10% or 1% of individuals from the population (third and fourth lines down, respectively) is used to reconstruct a lineages-through-time plot, the plot becomes shallower rather than steeper towards the present.

nus death rate) while the slope in the very recent past is the actual birth rate. It follows that it is possible to estimate both the population birth rate and the population death rate in the absence of explicit historical data. Indeed, maximum likelihood estimators have been produced and applied to real data sets, although the latter constitute species in a clade rather than individuals in a population (Nee *et al.* 1994a,b).

5.2 SAMPLING A SMALL PROPORTION OF THE POPULATION

In practice, ecologists and conservation biologists will not obtain sequence data to reconstruct a family tree relating all the individuals in a population. Rather, at best, they are likely to get data from a small number of individuals that constitute a fraction of the total population. If, for the moment, we restrict our consideration to a constant-rates birth-death process, it is possible to get insight into what happens to the shape of a lineages-through-time plot when only a small proportion of individuals from a local population has been sampled (Fig. 5.3). Intuitively, more recent nodes are likely to be disproportionately underrepresented in a genealogy constructed from a sample of individuals. Consider a full genealogy. Ancient nodes tend to give rise to more contemporary individuals than do recent nodes: for example, in a bifurcating tree the root node gives rise to all contemporary individuals, whereas the most recent node gives rise to just two. If we take a small sample of individuals from a population, it is unlikely that both descendants from the youngest node will be represented, but it is likely that the sample will contain at least one individual descended from each of the daughter lineages emanating from the most ancient node. Following this line of reasoning for nodes of intermediate age, and remembering that each daughter lineage has to have left at least one contemporary descendant for a node to appear, it is obvious why the more recent nodes will tend to be disproportionately underrepresented in the sample genealogy. It is not then surprising, as is evident from Figure 5.3, that the semilogarithmic lineages-through-time plot for smaller samples becomes shallower towards the present. Unfortunately, this means that the steepening towards the present that allowed us to estimate per lineage birth and death rates in Figure 5.2, has been eliminated; and parameter estimation is rendered much more difficult. However, by applying coalescence theory to estimate the expected distribution of sequential coalescence times going from the present to the root of the tree, even though it has not yet proved possible to estimate parameters with any accuracy (see below), it is possible to distinguish among different scenarios for population history (Nee *et al.* 1995). Figure 5.4 shows how this can be achieved by sequentially applying different transformations to the number of lineages or the time axis.

The procedure in Figure 5.4 has been applied to a number of case studies outside conservation biology (e.g., Holmes *et al.* 1995, 1996). For example, for viruses it revealed a Hepatitis C virus pandemic, which started about fifty years ago, and a Dengue fever virus endemic, for which the exponent has been increasing with time. In Figure 5.5 we give an example, specifically related to conservation biology, which focuses on the problems with using this approach given currently available material.

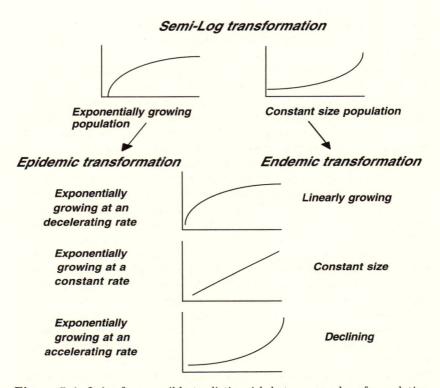

Figure 5.4. It is often possible to distinguish between modes of population growth using lineages-through-time plots. Nee *et al.* (1995) suggest examining the curvature of the semilogarithmic lineages-through-time plots to help distinguish among the various possibilities. A plot that becomes shallower towards the present is expected from an exponentially growing population, whereas one that steepens towards the present is expected from a constant-sized or declining population. It is then possible to apply appropriate transformations given by Nee *et al.* (1995) to the number-of-lineages axis (the epidemic transformation) or to the time axis (the endemic transformation) to further refine the appropriate model. Rambaut *et al.* (1997) provide a computer application that imports phylogenetic trees or gene genealogies and performs the appropriate graphical and statistical analyses.

5.3 PROBLEMS, ASSUMPTIONS AND UNCERTAINTIES

Neutrality

The models for the transformations used in Figure 5.4 are based on neutral genetics. Selection complicates matters. For example, selective sweeps that occur when a favourable mutation rapidly spreads through a population will eliminate standing variability from an asexual population. If a selective sweep occurred in a population that had otherwise been evolving according to neutral genetics, any current variability would have arisen since the origin of the mutant that would determine the deepest possible root of the genealogy; it would not be possible using that evidence alone to determine if the population had been through a recent extreme bottleneck or whether there had been a selective sweep. In one sense the difference between a selective sweep and a bottleneck is illusory: in both cases present-day individuals are descended from one or a few individuals at some specified time in the past. The difference is that with a selective sweep, the population size at that time may have been large but with a bottleneck it was small.

If occasional selected lineages have not reached fixation, statistical methods are available to distinguish parts of the genealogy that have been growing at a faster rate than others (and might be considered separately); those methods are incorporated into the computer application END-EPI (Rambaut *et al.* 1997). That same computer application also imports phylogenetic trees or gene genealogies constructed with standard tree-building programs such as PHYLIP or PAUP and performs the appropriate graphical and statistical analyses described in Figure 5.4.

Recombination

In order to construct trees using divergence, we assume that lineages have retained their integrity through time. Any parts of the genome that recombine among lineages change the tree into a network, and thereby invalidate the particular analyses described here. Looking back in time from the present, we assume that lineages coalesce, but with recombination, when one looks back, a single lineage should appear to divide (it derives from two parent lineages) rather than coalesce. Accordingly, biologists have focused on the use of nonrecombining mitochondrial DNA for analyzing eukaryote genealogies and phylogenies. Nonrecombining RNA viruses are ideal material for the type of genealogical analyses described in this chapter, because long nucleotide sequences are available that evolve at such a rapid rate, particular the envelope gene, which is forever being selected to escape the attention of the host immune system.

When recombination occurs at a known low rate, it should be possible to develop methods for incorporating its effects into analyses aimed at revealing past population dynamic history. When recombination rates are higher, it may be possible to identify sufficiently small lengths of genome such that the probability of recombination within them is vanishingly small, and then combine the results from the analysis for each length of genome on the assumption that the individual genes have assorted independently in the past. In that case, each gene genealogy will represent an independent sample of individuals that existed in the past and, therefore, will provide an independent estimate of population history. The use of microsatellite data for such analyses is particularly promising (Goldstein and Schlotterer 1998). Further, novel analyses of other gene sequence data (such as Alu insertions), based on appropriate evolutionary models, can similarly be used to analyze population dynamic history (Sherry *et al.* 1997). At the moment, we are witnessing the publication of preliminary analyses using a variety of genomic constituents, and it will probably take some years before the most cost effective procedures are in place to make appropriate genetic sampling and analyses routine.

Time Scale

Returning to the humpback whale example, only short lengths of genome in the mitochondrial control region contain material that is ideally suited for population analysis — sites within them evolve at a reasonably rapid rate. Those short lengths, generally between 300 and 600 base pairs, contain relatively little phylogenetic information. For example, if we examine a tree attempting to reveal the relationships among human races, only about six bases have changed from the root of the tree to the present, which means that the relationships revealed are subject to considerable statistical error: many lineages differ only by one or two bases pairs, and then at sites that have particularly high mutation rates and at which back mutation cannot be ruled out. In the whale example given in Figure 5.5, the mitochondrial control region sequenced is less than 300 sequences, and less than 2% of bases change from the root of the tree to the tips, which is less than six bases changes. We may therefore have relatively little confidence in the topology of the tree (bootstrap values are relatively low), though it does seem reasonable to assume that accumulation of lineages through time in the tree is reasonably accurate. However, the inference about population history can only be made over the time scale of the coalescences, which, in the whale case, is measured

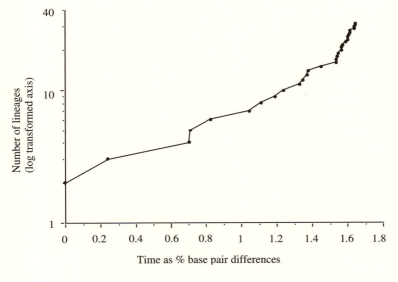

Figure 5.5a

Figure 5.5. Baker *et al.* (1993) report data on mitochondrial DNA control region sequences (283 base pairs) from 32 humpback whales (*Megaptera novaeangliae*) collected from around the world that differ by more than a single base pair. (a) The semilog representation of their phylogeny gives a plot steepening towards the present, suggesting that the endemic transformation is appropriate. (b) The internode intervals in their phylogeny exhibit linearity under the endemic transformation, suggesting they are consistent with having been drawn from a population of roughly constant size over the period of time covered by the coalescences. Because the time axis is formed by concatenating transformed intervals, there is no obvious scale for the axis, so none is shown. With this number of sequences, we do not expect the first coalescence to occur before, at the very least, about 10 generations ago (assuming that the whale population has been 5,000 since the cessation of harvesting). Hence, the impact of hunting is not expected to be manifested in this data. The phylogenetic tree used was reconstructed using the ultrametric KITSCH clustering program from the PHYLIP package (Felsenstein 1993). Similar trees were obtained using methods that allow variable rates of substitution. Distances were corrected for multiple substitution using a model of molecular evolution that allows different base frequencies and different rates of transition and transversion (PHYLIP program DNADIST). The transition/transversion ratio was set to 20:1. This example is taken from Nee *et al.* (1995).

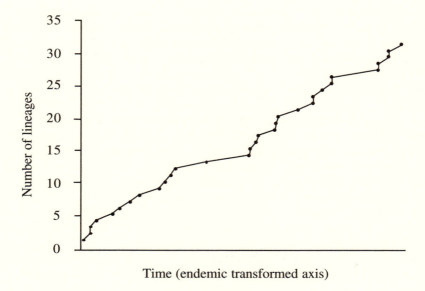

Figure 5.5b

over more than a million years (sequence divergence rate is estimated to 0.7% to 1% per million years) and not over the decades that would tell us whether there had been a recent population crash as a result of overwhaling by humans.

If we were attempting to determine the phylogenetic relationships among classes of vertebrates or orders of mammals, we should not use the control region because it evolves too rapidly, but we should need to examine the whole of the rest of the genome in order to produce a reliable phylogeny, which means sequencing all 16,000 sites (Cummings *et al.* 1995)!

An example from conservation biology of the importance of getting an accurate time scale is reported by Moritz (1995), who describes the gene genealogy of Pacific Ocean populations of the coconut crab, *Birgus latro*, as being starlike. The implication is that the logarithmically transformed lineages-through-time would approximate to that in the top left of Figure 5.4, suggesting an exponentially growing population. In fact, these populations of the coconut crab are currently going extinct! Presumably there had been rapid expansion of the crab populations at some time in the past, probably during the lower sea levels of the Pleistocene, which was the period over which the coalescences occurred. If many individuals in a sample are identical for the genetic sequence used to construct the genealogy, then it will be necessary to identify a sequence that evolves more rapidly and therefore carries the signature of more recent genetic change and coalescent events.

The Molecular Clock

One assumption which is frequently made in order to transform units of genetic distance to time is a molecular clock. If the clock assumption is not made and branches evolve at different rates, the tips will not line up and temporal analyses are impossible. It is possible to test for rate heterogeneity and, if the date accords with rate constancy, that assumption can be made in the tree generation program so that tips do line up and temporal analyses can be performed. For example, the analysis by Ong *et al.* (1997) of papillomavirus gene sequences could not reject the hypothesis of rate constancy, and they therefore used a maximum likelihood procedure for tree generation that assumed a constant rate of change. For cases where the hypothesis of rate constancy is rejected by the data, it is desirable to develop methods that incorporate rate heterogeneity while still allowing nodes to be placed in relative time. Andrew Rambaut and colleagues (pers. comm.) have proposed one way forward. They start by producing a maximum likelihood tree in which rates of molecular evolution are allowed to differ among all the branches, and they then use a likelihood ratio test to determine whether this tree differs significantly from a tree with the same topology constructed under the assumption of a single rate (Felsenstein 1981). If the single-rate tree is significantly worse, the point at which a single-rate change for a clade increases the likelihood most is determined, and the likelihood of the two-rate tree will then be compared with that of the multiple-rate tree. This procedure can be repeated until a tree is found that is not significantly worse than the multiple-rate tree. Under this procedure, given the known point(s) at which rates of evolution change, it will be possible to map nodes in relative time without the restrictive assumption of a single molecular clock. The method can be programmed and techniques can be developed to make it run at reasonable speeds on standard workstations.

Sampling

The analytical models on which population history analyses are produced assume phylogenetically random sampling of individuals from the contemporary population. In fact, at the moment, no data sets are available in which sampling has been designed to produce a random sample.

The extent to which nonrandom sampling changes the shape of lineages-through-time plots, or tree topology in general, is an unexplored area. As with many problems concerning tree structure, intuition frequently leads us astray. The only real advances will be made through modeling and computer-simulation exercises that are designed to mimic specific case studies as well as to reveal general principles. For the foreseeable future, the sampling problem is going to be evident with all comparative genome analyses (even though it may continue to be ignored). A useful property of the coalescence approach is that a phylogenetically random sample from a coalescent tree should produce an almost identically shaped lineages-through-time plot (there is some very slight departure under the epidemic transformation). This means that it should be possible to take nonrandom samples from a simulated coalescent tree and examine how those influence tree structure, measured both as tree balance and as the shape of a lineages-through-time plot. One procedure would be to produce a sample as follows. Choose and accept a random individual. Base the probability of accepting a second randomly chosen individual into the sample on its genetic relationship to the first individual. The probability of subsequent randomly chosen individuals being accepted into the sample is based on their genetic relationships to other individuals already accepted into the sample. If we wished to get phylogenetic underdispersion (mimicking, for example, the case of sampling from a geographically restricted group from a larger population) then close relatives would be more likely to be accepted into the sample. To get overdispersion (mimicking, for example, the case where a single representative sample is taken from each geographical area covered by the distribution of a population) no close relatives would be accepted. The degree to which sampling has been phylogenetically nonrandom is reasonably well measured by total branch length in the reconstructed tree: phylogenetically underdispersed sampling results in a short total branch length, while phylogenetically overdispersed sampling results in an unexpectedly long total branch length. The whale data analyzed in Figure 5.5 are likely to be reasonably phylogenetically random because samples were taken from the world range of the species, but sampling was not structured so that only a single representative was taken from each area.

Parameter Estimation

While it may be possible to reveal the qualitative nature of a population's dynamic history using the method described in Figure 5.4, it is more difficult to estimate parameter values from gene genealogies. Unfortunately, many parameters describing tree structure are composites of component parameters of interest (population sizes, birth rates, death rates, mutation rates, generation time, and real time). We can, in fact, estimate some of these parameters from other sources, and the time is approaching when we can use those estimates together with composite parameter values estimated from trees to estimate other parameters of interest.

Currently, there are two main approaches for estimating population genetic parameters from gene sequence data. The first is to use the genealogical structure relating sampled sequences (Felsenstein 1992; Fu 1994; Fu and Li 1993; Griffiths and Tavare 1994a,b; Kuhner *et al.* 1995; and Lundstrom *et al.* 1992). The second approach is to use summary statistics, such as the number of segregating sites or the distribution of pairwise differences (Griffiths and Tavare 1996; Rogers and Harpending 1992; Tajima 1983; Watterson 1975). The first approach has the advantage that, because it makes full use of the available data, parameter estimates have smaller variances (Felsenstein 1992; Lundstrom *et al.* 1992). In contrast, the second approach can be computationally much faster and allow the implementation of more complex population dynamic and mutational models.

The method used by Rannala and Yang (1996) for estimating tree structure under the assumption of an explicit birth-death process suggests one promising way forward using the genealogical structure approach. The tree with the highest posterior probability is chosen as the most likely, and maximum likelihood estimates of birth and death rates are estimated explicitly as part of the tree estimation process. Currently, their method is designed to deal only with complete trees and not small samples of extant lineages. In principle, however, there is no reason why it cannot be adapted to deal with small samples. It should ultimately be possible to develop tree-based methods for estimating $2uN_0$, $2uN_1$, and $2ut$, where N_0 and N_1 denote the effective population sizes at the beginning and end of the period of the coalescence, u is the aggregate mutation rate over the region of DNA under study, and t is the number of generations over the period of the coalescence. Given knowledge of some parameters such as the current population size and the mutation rate, the three composite parameters could then be used, for example, to estimate the population growth rate.

Alternative Methods

The methods based on summary statistics do not generally progress through a tree-building stage. One potential problem here derives from the fact that there are not as many statistically independent comparisons as there are pairwise gene sequence comparisons. For example, for most trees well over one-third of comparisons between individuals are in fact different estimates of divergence arising from the root of the tree (Felsenstein 1992). Nevertheless, simulation studies are now frequently performed under different scenarios of population change to provide distributions of pairwise differences without an explicit tree-building stage, and statistical comparisons can be made between those distributions with the one found in any particular data set (e.g., Rogers 1995; Rogers *et al.* 1996). Indeed, analytical approximations have been derived for estimating population growth parameters from the distribution of sequence distance comparisons (Rogers and Harpending 1992; Weiss *et al.* 1996). Even here, most comparisons concern the few very deepest nodes, whereas our interest as conservation biologists is likely to be in the more recent nodes than the ancient nodes. That being said, it is always possible to argue that because pairwise comparison analyses do not include an explicit tree-building phase, they are to be preferred because they do not need to deal with the uncertainties of tree reconstruction (Rogers *et al.* 1996). Felsenstein (1992) demonstrated that under a scenario of constant population size, the statistical power of pairwise methods is weak compared with a maximum likelihood method based on phylogenetic estimates. Whether the same is true for growing populations is strongly contested by Rogers *et al.* (1996), despite an elegant analysis by Bertorelle and Slatkin (1995), who demonstrate, with a simulation study utilizing the number of segregating sites, that pairwise methods appear to estimate demographic parameters incorrectly. There is an urgent need to compare the statistical power of methods that proceed through the tree-building stage with those that do not. If these studies are not performed, the two very different approaches to data analysis will each continue to develop as though the other did not exist.

5.4 CONCLUSIONS

It would be a reasonable accusation that the contents of this chapter have told us nothing about conservation biology: the two examples relating to conservation, the humpback whale and the coconut crab, dealt with evolutionary rather than ecological time scales. The defense is that conservation biologists need to know the dynamic history of populations but usually have very little if any census data from the past. Genealogical analyses of the type suggested here potentially provide a means for getting information about population dynamic history. That information is encoded in the genetic material, and we are now beginning to understand how to decode it. At the moment, as we have seen, epidemiologists are revealing new information about the spread of virus diseases. The time is ripe for conservation biologists to reveal new information about the dynamic status of the populations they wish to conserve.

REFERENCES

Baker, C.S., Perry, A., Bannister, J.L., Weinrich, M.T., Abernethy, R.B., Calambokidis, J., *et al.* (1993). Abundant mitochondrial DNA variation and world-wide population structure in humpback whales. *Proc. Natl. Acad. Sci. USA*, *90*, 8239–8243.

Bertorelle, G., and Slatkin, M. (1995). The number of segregating sites in expanding human populations with implications for estimates of demographic parameters. *Mol. Biol. Evol.*, *12*, 887–892.

Cummings, M., Otto, S. and Wakeley, J. (1995). Sampling properties of DNA sequence data in phylogenetic analysis. *Mol. Biol. Evol.*, *12*, 814–822.

Felsenstein, J. (1981). Evolutionary trees from DNA sequences: a maximum likelihood approach. *J. of Mol. Evol.*, *17*, 368–376.

Felsenstein, J. (1992). Estimating effective population size from samples of sequences: inefficiency of pairwise and segregating sites as compared to phylogenetic estimates. *Gen. Res., Cambridge*, *59*, 139–147.

Felsenstein, J. (1993). *PHYLIP (Phylogeny Inference Package). Version 3.5c.* Distributed by Author at Department of Genetics, University of Washington, Seattle, Washington 98195, U.S.A.

Fu, Y.X. (1994). A phylogenetic estimator of effective population size or mutation rate. *Genetics*, *136*, 685–692.

Fu, Y.X. and Li, W.H. (1993). Maximum likelihood estimation of population parameters, *Genetics*, *134*, 1261–1270.

Goldstein, D.B. and Schlotterer, C., eds. (1998). *Microsatellites: Evolution and Applications.* Oxford University Press, Oxford.

Griffiths, R.C. and Tavaré, S. (1994a). Ancestral inference in population genetics. *Stat.Sci.*, *9*, 307–319.

Griffiths, R.C. and Tavaré, S. (1994b). Sampling theory for neutral alleles in a varying environment. *Phil. Trans. Roy. Soc. Lond. B*, *344*, 403–410.

Griffiths, R.C. and Tavaré, S. (1996). Monte Carlo inference methods in population genetics. *Math. Comput. Mod.*, *23*, 141–158.

Harvey, P.H. (1995). The WORLDMAP debate: reply from P.H. Harvey. *Trends in Ecology and Evolution*, *10*, 82–83.

Harvey, P.H., May, R.M., and Nee, S. (1994). Phylogenies without fossils. *Evolution*, *48*, 523–529.

Holmes, E.C., Bollyky, P.L., Nee, S., Rambaut, A., Garnett, G.P., and Harvey, P.H. (1996). Using phylogenetic trees to reconstruct the history of infectious epidemics. In Harvey, P.H., Leigh-Brown, A. J., Maynard Smith, J. and Nee, S. (eds.), *New Uses for New Phylogenies*, 169–186, Oxford: Oxford University Press.

Holmes, E.C., Nee, S., Rambaut, A., Garnett, G.P., and Harvey, P.H. (1995). Revealing the history of infectious disease epidemics using phylogenetic trees. *Phil. Trans. R. Soc. Lond. B*, *349*, 33–40.

Kuhner, M.K., Yamato, J., and Felsenstein, J. (1995). Estimating effective population size and mutation rate from sequence data using Metropolis-Hastings sampling. *Genetics*, *140*, 1421–1430.

Lundstrom, R., Tavaré, S., and Ward, R.H. (1992). Estimating substitution rates from molecular data using the coalescent. *Proc. Natl. Acad. Sci. USA*, *89*, 5961–5965.

Moritz, C. (1995). Uses of molecular phylogenies for conservation. *Phil. Trans. R. Soc. Lond. B*, *349*, 113–118.

Nee, S., Holmes, E.C., May, R.M., and Harvey, P.H. (1994a). Extinction rates can be estimated from molecular phylogenies. *Phil. Trans. R. Soc. Lond. B*, *344*, 77–82.

Nee, S. and May, R.M. (1997). Extinction and the loss of evolutionary history. *Science*, *278*, 692–694.

Nee, S., May, R.M., and Harvey, P.H. (1994b). The reconstructed evolutionary process. *Phil. Trans. R. Soc. Lond. B*, *344*, 305–311.

Nee, S., Holmes, E.C., Rambaut, A., and Harvey, P.H. (1995). Inferring population history from molecular phylogenies. *Phil. Trans. R. Soc. Lond. B*, *349*, 25–31.

Ong, C.-K., Nee, S., Rambaut, A, Bernard, H.-U., and Harvey, P.H. (1997). Elucidating the population histories and transmission dynamics of Papillomaviruses using phylogenetic trees. *J. Mol. Evol. 44*, 199–206.

Rambaut, A., Grassly, N.C., Nee, S., and Harvey, P.H. (1996). Bi-De: an application for simulating phylogenetic trees. *Comput. Appl. for Biosci. 12*, 469–471.

Rambaut, A., Harvey, P.H., and Nee, S. (1997). End-Epi: an application for reconstructing phylogenetic and population processes from molecular sequences. *Comput. Appl. Biosci. 13*, 303–306.

Rannala, B. and Yang, Z. (1996). Probability distribution of molecular evolutionary trees: a new method for phylogenetic inference. *J. Mol. Evol. 43*, 304–311.

Rogers, A.R. (1995). Genetic evidence for a Pleistocene population explosion. *Evolution*, *49*, 608–615.

Rogers, A.R., Fraley, A.E., Bamshad, M.J., Watkins, W.S., and Jorde, L.B. (1996). Mitochondrial mismatch analysis is insensitive to the mutational process. *Mol. Biol. Evol. 13*, 895–902.

Rogers, A.R., and Harpending, H.C. (1992). Population growth makes waves in the distribution of pairwise genetic differences. *Mol. Biol. Evol.*, *9*, 552–569.

Sherry, S.T., Harpending, H.C., Batzer, M.A., and Stoneking, M. (1997). *Alu* evolution in human populations: using the coalescent to estimate effective population size. *Genetics*, *147*, 1977–1982.

Tajima, F. (1983). Evolutionary relationships of DNA sequences in finite populations. *Genetics*, *105*, 437–460.

Vane-Wright, R.I., Humphries, C.J., and Williams, P.H. (1991). What to protect? — Systematics and the agony of choice. *Biol. Conserv.*, *55*, 235–254.

Watterson, G.A. (1975). On the number of segregating sites in genetical models without recombination. *Theor. Popul. Biol.*, *7*, 256–276.

Weiss, G., Henking, A., and von Haesler, A. (1996). Distribution of pairwise differences in growing populations. In Donnelly, P. and Tavaré, S. (eds.), *Progress in Population Genetics and Human Evolution*. Berlin: Springer-Verlag.

Williams, P.H., Gaston, K.J., and Humphries, C.J. (1994). Do conservationists and molecular biologists value differences between organisms in the same way? *Biodiversity Lett.*, *2*, 67–78.

6

Parasites and Conservation of Hawaiian Birds

REBECCA L. CANN AND LESLIE J. DOUGLAS

SUMMARY. Conservation geneticists can build on the experience of specialists from many disciplines, and this chapter discusses the importance of malaria in the extinction of native Hawaiian birds. Epidemiologists often consider the ease of transportation, the local occurrence of water sources, and the number of naive individuals in a population as prime determinants in the spread of infectious diseases. As in human epidemiology, all these factors appear to have combined in Hawaii, leading to extinction and even wholesale eradication of our lowland tropical bird faunas. Before the development of sensitive molecular diagnostic techniques, conservation biologists grossly underestimated the magnitude of this effect with avian malaria, a devastating infectious disease in wild bird populations. Indeed, without accurate identification of infected individuals, especially those birds with low parasite levels, wildlife biologists could downplay the importance of alien species as reservoirs, the evolution of disease resistance, and the effectiveness of different ex-situ management actions aimed at increasing population size. Instead, they focused on eradicating competitors, excluding predators, and protecting habitat; however, in spite of these actions, bird populations have not automatically recovered. Better molecular tools have led to increased awareness of the true scope of the problems that face conservation biologists, and these tools also offer some hope for the identification and selective rearing of genotypically resistant individuals in environments where the disease cannot be eradicated.

INTRODUCTION

Diseases that vary in severity between members of a population are potent evolutionary forces (Ewald 1994). They promote the long-term maintenance of genetic variation and result in advantageous arrangements of genomic sequences for quick activation of host defenses, as well as assist in the rapid replacement of individual alleles within and between populations (Hamilton 1980). In this sense, a disease epidemic can leave its mark on the genome of a species long after a particular cycle of infection and immunity has been resolved.

From a broader evolutionary perspective, it is also important to ask how often diseases and parasites are linked to extinction. Some notable vertebrate and invertebrate examples, such as gorillas with respiratory illnesses, harbor seals, lions, and ferrets attacked by viruses, parrots infected with botflies, and crayfish besieged by fungi have been well-documented (Thorne and Williams 1988; Dobson and Miller 1989; Loye and Carroll 1995). However, some conservation biologists have been slow to incorporate the perspective that disease offers to their thinking, perhaps because they have a plethora of ready examples in organisms threatened by immediate habitat loss, and judge its importance at a lower level of priority.

Unfortunately for native Hawaiian taxa, problems with introduced diseases, predators, and habitat loss come together in the same package. Effective management of remnant populations now depends on a clear understanding of host-parasite interactions and genetic diversity in fragmented populations, as well as fencing, surveillance, captive-rearing, and translocation (Freed and Cann 1989). Molecular approaches to conservation can help provide a bridge between practical managment strategies and theoretical population models. More importantly, molecular techniques can supply evidence that current strategies often force endangered species into high-density populations that speed the transmission of infectious diseases (Scott 1988), producing consequences that require further attention.

Our aim is to illustrate these points using examples of avian malaria, and discuss its role in the extinction of Hawaiian forest birds. We will try to (1) update nonspecialists in the progress made since the identification by Warner (1968) of malaria as the most important disease limiting native bird populations, (2) estimate its overall prevalence, and (3) report the discovery of what we suspect may be a resistant/tolerant population of native birds. The goals of preservation and recovery often seem elusive because of fundamental gaps in our knowledge about the natural history of target species, but the stakes are very high. If we want to prevent the Hawaiian archipelago from providing yet another illustration of an oceanic avifauna approachable only from a paleontological perspective (Steadman 1989), new tools must be employed.

6.1 DISEASE IN HAWAIIAN BIRDS

It is a truism of parasitology that parasites exist to exploit nonequilibrium conditions (Price 1980). They are usually highly localized in space and may pass through an infective phase that is often brief in time. One might expect that a population of intracellular parasites would be difficult to establish in Hawaii, where only the most healthy individuals survive the hardship of long-distance dispersal. However, volcanic activity also controls weather patterns that affect movement of hosts and parasites by both wind and sea, and rapidly fluctuating habitat patches are characteristic of Hawaiian ecosystems (Moore *et al.* 1994). Survival of parasites under these conditions might not be expected to depend so much on genotype as much as on chance.

Humans gave some parasites an enhanced chance in 1826, when they accidentally imported mosquitoes to Maui and set in place conditions for the effective transmission of haematozoa by a bloodsucking vector (Atkinson and van Riper 1991). The historical importance of diseases in Hawaiian bird populations was suspected by the early naturalists Henshaw (1902) and Perkins (1903), who encountered dead and dying birds in the forest, but evidence for epizootic transmission was first recognized and stated systematically by Warner (1968), who worked during a period when the reasons for the disappearance of native birds were poorly understood. Alien passerine species that potentially outcompeted natives were actively introduced to the islands as late as the 1960s, alien predators were abundant, forest habitats were rapidly destroyed or degraded as human populations grew, and resort development was unrestrained (Pimm 1991). Prior to Warner's study, all these forces were thought to be more important than introduced diseases. Warner's observations that most Hawaiian birds appeared to be confined to high-elevation habitats, that they appeared to be especially susceptible to malaria when deliberately infected with an experimental challenge compared to introduced species, and that their primary distribution was negatively correlated with abundant mosquito populations in the lowlands provoked the first serious evaluation of disease as a factor in the extinction of Hawaiian birds.

Almost twenty years later, the enormity of this pattern is clearly evident on all the main islands, each with its own unique history of habitat loss, alien introductions, and resource mismanagement. The island-by-island census of bird population densities, tabulated as the Hawaiian Forest Bird Survey of Scott *et al.* (1986), served to identify steep elevational patterns and latitudinal gradients that characterized highest bird population densities on every island with the exception of Oahu and Lanai. Native forest birds were largely restricted to specific elevations above 1500–1800mm as predicted by Warner. A corollary of the field surveys was the documentation of apparently pristine, intact forests at lower elevations in which native birds were missing. These were forests

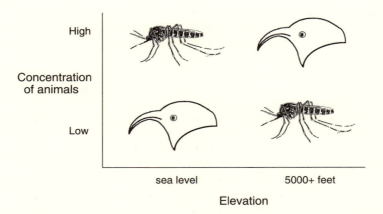

Figure 6.1. This figure shows the inverse relationship between concentrations of native birds and introduced mosquitoes, defining the highest priority habitats where conservation efforts are directed at saving endangered birds in Hawaii today.

with few competitors or predators, and that had habitats with appropriate food and/or nesting resources for a variety of endangered and extinct species. They are also some of the forests with the greatest densities of mosquitoes. (Leonard Freed discusses these forests in Chapter 7 in this volume.)

Over nineteen orders and fifty families of avian parasites have been identified in Hawaii so far (van Riper and van Riper 1985). At least eight species of mosquitoes have been introduced to Hawaii (Goff and C. van Riper 1980), two of them for biocontrol purposes, and five are known to take blood meals, but the principle vector of avian malaria in Hawaii is now considered to be a single Culex mosquito, *Culex quinquefasciatus* (Laird and C. van Riper 1980). Six species of *Plasmodium* have also been reported in Hawaiian birds, and at least three seem to have been the source of major epizootics (*P. relictum, P. gallinaceum, and P. elongatum*). The mosquito bites primarily at night, and a single bite from an infected mosquito is sufficient to kill some native birds (Atkinson *et al.* 1995).

Although densities are not especially high in all forests, virtually all of the *C. quinquefasciatus* mosquitoes trapped have been found to carry *Plasmodium,* and the numbers of parasites in a single mosquito are some of the most extreme known to entomologists (D. La Ponte, pers. comm.). Two bites can result in 100% mortality, using specific populations of the native 'Apapane (*Himatione sanguinea*) and the Common 'Amakihi

(*Hemignathus virens*). Surprisingly, the parasite strains found in Hawaii are not especially virulent, when compared to those circulating in forests on the U.S. mainland (C. van Riper III, pers. comm.). Knowledge of parasite species and strains is not based on DNA or RNA sequence identification so far. *P. relictum capistranoae* is thought to be the primary parasite driving epizootics in Hawaiian forest birds today, based on characteristics of the pathogen found by using light microscopy. Charles van Riper, Carter Atkinson, and Lee Goff have followed malarial parasites and their vectors for over twenty years, characterizing the elevation gradients that mark their distributions on different islands.

Perhaps just as importantly, these researchers have also identified that after a single bite (which in the native I'iwi usually results in less than 10% survival rate), all birds exhibit changed behaviors. This illustrates that in addition to evolving biological defenses to parasites, Hawaiian birds are also capable of undergoing behavioral shifts. Warner originally thought that Hawaiian birds from low-elevation habitats had a peculiar sleeping posture that exposed their feet and beaks to mosquito bites. It is known that when canaries are artificially infected with *P. relictum*, they show a decreased thermoregulatory capacity (Hayworth *et al.* 1987). Thus, native birds could potentially undergo behavioral shifts that might help them cope with the presence of mosquitoes by altering their roosting and resting postures. These changes would be especially important to their survival in the lowlands. In addition, such behavioral alterations might also help them adapt to temperature extremes in high-elevation rain forests, which may not be their optimum habitat.

One compensatory behavior of Hawaiian passerines that had been proposed was a diurnal movement pattern between low-elevation forests for foraging and high-elevation refuges for sleeping (C. van Riper *et al.* 1986). Under these conditions, infections acquired during feeding bouts in low-elevation forests are likely to be self-limiting, because infected individuals would probably not survive the night in forests with freezing temperatures each month of the year, only to be bitten by the rare mosquito that makes its way up the volcano on a warm air current. We have found both native and introduced birds infected with *Plasmodium* above 5,000 feet, but so far epizootics in these habitats are probably rare (Feldman *et al.* 1995). Global warming trends will no doubt alter this pattern in the years to come.

6.2 THE ROLE OF INTRODUCED SPECIES

An additional factor contributing to the maintenance of infectious disease cycles in Hawaiian forests is the presence of very large populations of

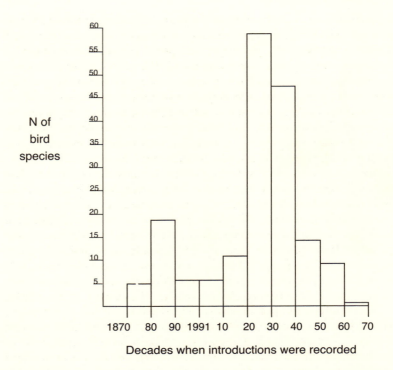

Figure 6.2. This figure indicates alien bird introductions by decade to the six largest Hawaiian islands (after Mouton and Pimm 1983). Some species are represented by multiple introductions, because the original attempt was unsuccessful. The peak of introductions occurred between 1920 and 1940. Bird species that were represented by only single pairs were not even counted in this analysis.

introduced avian species. Deliberate introductions began in earnest after 1870 (Fig. 6.2), and continue today. Pratt (1994) states that more avian species have been introduced into the Hawaiian Islands than anywhere else on earth. At least 125 species introductions have been recorded, and one estimate (Berger 1973) is a minimum of 156, beginning with chickens introduced by Polynesians circa 200 C.E., followed by the introduction in 1796 of pigeons (*Columba livia*).

These facts become important in the following context. When a few alien birds were tested for their ability to survive deliberate infections with *Plasmodium*, Charles van Riper and colleagues (van Riper *et al.* 1986) discovered that introduced species such as the Red-billed Leothrix (*Leiothrix lutea*), the Japanese White-Eye (*Zosterops japonicus*), and a canary (*Serinus canarius*) either had very low levels of parasitemia or none that were measurable, as judged by blood smears under challenge conditions.

Birds from temperate and tropical continents evolved with the presence of malarial parasites in their ecosystems, while the Hawaiian forest birds did not. Most of the bird species brought into Hawaii have been there for at least fifty generations, and we should expect that there may have also been microevolutionary changes in their host-parasite interactions, compared to ancestral lineages. The failure, at first, to discover large numbers of parasites in introduced species led some biologists to (1) discount the role these aliens have in epizootic transmission among natives; and (2) look instead for more subtle ways in which introduced birds might be competing with natives for nesting opportunities and/or food. By focusing attention on some of the most recently introduced species (two species of bulbuls came into Hawaii in the 1960s), we wondered if we could gain a clearer understanding of the dynamics of disease transmission among the birds.

Details of our new diagnostic for avian malaria have been previously published, along with a comparison of sensitivity compared to traditional smear technology and an estimate of thresholds of sensitivity for detection of parasites at very low levels of infection (Feldman *et al.* 1995). The test targets a region of the small nuclear ribosomal RNA gene (18S rRNA), which varies in size and sequence between parasites and hosts, has internal controls for amplification efficiency, has positive and negative controls (Fig. 6.3), and offers at last the promise of fast and accurate malarial detection for about $1.20 per bird sample, including the cost of preparing the DNA. Birds are caught in aerial and pole-based mist nets and are then banded, weighed, bled, measured, photographed, and released. The 50-microliter blood sample taken from the brachial vein contains the genome of the bird, along with the genomes of blood-borne infectious agents.

First, this test documents the presence of malaria in introduced birds and endangered species in our wildlife refuges and natural area preserves on Maui and Hawaii, which were thought to be safe habitats because these forests were above 5,000 feet in elevation. What a dangerous assumption! Both native and introduced species contain a few positive individuals.

Second, we employed this test to look for the presence of *Plasmodium* in a variety of habitats, and have uncovered evidence that a large reservoir population of parasites is potentially maintained in low-, mid-, and upper-elevation forests by at least nine species of introduced birds (Shehata *et al.* in press; Cann *et al.* 1996). Infected birds at low elevations on Oahu showed an average species prevalence level of about 10%

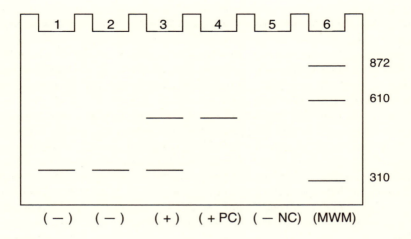

Figure 6.3. A cartoon of an agarose electrophoretic gel. In this cartoon, the 6 lanes show the products of a PCR reaction separated by size. A portion of the 18S rRNA gene has been amplified, using conserved primers that span a region that varies in length between vertebrates and *Plasmodium*, as in Feldman *et al.* (1995). Lanes 1 and 2 show uninfected birds, and have only a band that corresponds to the bird's own 18S rRNA fragment, while lane 3 shows a bird that has the same sized band, but also a second, higher molecular weight band that corresponds to the 18S rRNA fragment from *Plasmodium*. Lane 4 contains a positive control from purified *P. gallinaceum* DNA, while lane 5 is a negative control, reagents with no DNA. Lane 6 shows molecular weight markers that allow us to estimate the size of the amplified DNA products.

over a three-year period of monitoring, and at least five alien species had prevalence levels that exceeded 20%. The five introduced species with the highest prevalence rates were the white-rumped shama (*Cosychus malabaricus*), red-whiskered bulbul (*Pycnonotus cafer*), red-vented bulbul (*Pycnonotus jocosus*), nutmeg mannikin (*Lonchura punctulata*), and spotted dove (*Streptopelia chinensis*). We also documented that five introduced species sometimes suspected as reservoirs were free of malaria at the time of testing. These included common mynas (*Acridotheres tristis*), house finches (*Carpodacus mexicanus*), java sparrows (*Padda oryzivora*), northern cardinals (*Cardinalis cardinalis*), and zebra doves (*Geopelia striata*).

Our ability to detect as few as 8 parasites per 10,000 bird red-blood cells is essential, because of the expectation that many introduced species would show very low levels of parasites in their circulatory system. The test appears to be capable of detecting evidence of *Plasmodium* for up to a year after the initial infection event, judging from challenge experiments and subsequent monitoring of survivors via PCR (C. Atkinson, pers. comm.). Numbers of truly immune individuals will be more difficult to estimate and will require more sensitive Enzyme-Linked Immunosorbent Assay (ELISA) protocols to assay antibody titers than are available to us today.

We agree that under some circumstances in which native birds remain abundant, they are fully capable of acting as their own disease reservoir if a forest has been degraded by feral pig damage. Rooting pigs create hollows in the ground and in tree-fern logs. In the rainforest, these can become stagnant pools used by the vector for oviposition. Birds in such forests often show malarial prevalence rates of 20%–50% infection (Atkinson *et al.* 1995; Cann *et al.* 1996). Political and ethical arguments concerning the removal of nonnative pigs limit the ability of conservation biologists to break the cycle of malarial transmission in many Hawaiian forests containing native birds.

However, in forests where introduced species far outnumber native birds, the same cycle of pigs, mosquitoes, and parasites depends on a ready supply of hosts, even if individual parasites do not do as well. Here, the balance can shift, virulence might increase, and reestablishment of locally extirpated natives may depend on the ability of endangered species to tolerate or resist *Plasmodium*, which will always be present. Our efforts are now devoted to identifying more virulent parasite strains, characterizing resistant bird lineages, and discovering the molecular basis of these phenotypes.

6.3 DISEASE AND CONSERVATION PRIORITIES

How malaria limits conservation efforts directed at saving native Hawaiian birds is a complicated story. On the one hand, some long-term restoration attempts might be put on hold because certain main islands have no, or little, appropriate habitat over even 2,000 feet in elevation for forest birds (Table 6.1). When remaining habitats on these low islands are restored to native vegetation and predators are removed, the habitats are still incapable of supporting the flagship native bird species, because of the continual threat of malaria. Ecotourists and hikers are disappointed that they cannot see the famous birds, ignoring the plants and arthropods that supported their existence, and we lose unique opportunities for conservation education, because such sites are often adjacent to major urban recreational destinations.

Island	Percentage of Area
Hawaii	68.4
Maui	41.4
Kauai	24.0
Molokai	17.8
Lanai	6.3
Oahu	4.6
Kahoolawe	0.0
Niihau	0.0

Table 6.1. Percentage of island's land area at an elevation higher than 2,000 feet. Data from Armstrong (1973), p.204.

Older islands have eroded and are disappearing into the ocean, carrying unique bird lineages with them. This fact of geology creates a problem for forest-bird conservation on Lanai, Oahu, Molokai, and Niihau, as well as the relict populations of passerines on the northwest islands and atolls in the Hawaiian chain (especiallly Nihoa, Necker, Laysan, Pearl and Hermes Reefs, and French Frigate Shoals). High-elevation habitats judged most capable of sustaining forest-bird populations exist only on Maui and the island of Hawaii, and these islands receive the bulk of resources and research effort. However, our data tell us that complacency is dangerous, because malaria can be found above 5,000 feet on both Maui and Hawaii.

A second point that must be made is the impact, or lack thereof, that information about malaria has had on our captive rearing programs. Dedicated personnel in zoos and private agencies have been attracted to gather basic information about natural history and development of Hawaiian birds, and have hand-reared chicks from eggs, only to have these birds die of malaria when released as fledglings back into the wild (Kuehler *et al.* 1996). Could this have been avoided? Knowledge that certain 'Amakihi populations varied in their ability to resist and survive malarial challenges (van Riper *et al.* 1986) was ignored in the selection of eggs for rearing. Fortunately, biologists learned from this mistake and chose next a subpopulation of nonendangered native thrush, or 'Omao (*Myadestes obscurus*), from an area where malaria is endemic for their later reintroduction experiments, in preparation for work with endangered thrush on the island of Kauai.

Over 48 alien bird species have well-established breeding populations in Hawaiian forests, (Berger 1973), increasing the likelihood of continuous, virulent disease transmission. Most of these alien species now enjoy the protection of state wildlife officials, making it illegal to trap and remove them without permits! If disease-resistant or disease-tolerant genetic lineages of native birds coexist among them, it is imperative that we identify them and designate these individuals to be the progenitors of captive stocks. Such knowledge is also the key to reestablishing native bird diversity in low-elevation forests on Oahu, Molokai, and Kauai, where island-wide vector control is impossible, and where the introduction of alien species continues.

6.4 A LESSON FROM OAHU

The island of Oahu is often used to illustrate the extent of habitat modification in Hawaii, because while it gives the appearance to the casual visitor of a green, tropical paradise, very few native plant and animal species remain. Those managing to hold their own exist in fragmented patches on two low mountain ranges, the Waianae and Koolau. A deforested plateau approximately fifteen miles across, with vast expanses of monocultured crop plants and/or suburban housing, separates these ranges. Microevolutionary forces appear to be actively at work, and we expected to see some evidence of them in the genetic population structure of native birds.

We have used a conserved set of mitochondrial DNA primers to amplify a portion of the cytochrome B gene (Kocher *et al.* 1989), in order to assess how genetically variable the apparently disease-resistant Oahu 'Amakihi are, compared to populations drawn from other islands where 'Amakihi are also found. One point of interest is the relationships between Oahu lineages and Hawaii lineages, where there is known variation in survival rates upon malarial challenge. A quick summary of this information is seen in Figure 6.4, based on 250 nucleotides of sequence aligned from 68 birds. These diagrams show that the Oahu 'Amakihi lineages are evolutionarily divergent in their mitochondrial sequences compared to those lineages found on other islands. Due to a current lack of gene flow between islands, recent mutations that result in Oahu populations tolerating malaria may not be shared with other islands. However, all these lineages share a coalescent that is less than 1.2 million years old, based on a crude assumption of a standard molecular clock for mitochondrial coding regions. The short time scale seen in Oahu 'Amakihi lineages has important implications for other species.

Only four species of native passerines exist on Oahu in populations of more than 200 individuals. When Williams (1987) tabulated recent Christmas Bird Counts of the Audubon Society, he noticed that the continued decline of native species corresponded to a period when two

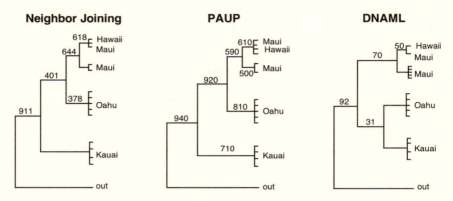

Figure 6.4. This figure shows the relationships of five separate mtDNA haplotypes identified from Oahu 'Amakihi, compared to 'Amakihi populations on other islands, using three different phylogenetic methods (distance, parsimony, and maximum likelihood). The designated outgroup was a honeycreeper from Kauai, the anianiau (*Hemignathus parvus*). The parsimony tree was constructed using PAUP 3.1.1 of Swofford. The distance tree (NJ) and the maximum likelihood tree (DNAML) were constructed using Felsenstein's PHYLIP package. Numbers on branches reflect bootstrap values for 1000 (NJ, PAUP) or 100 (DNAML) replications.

recently introduced birds were rapidly expanding in population size. At the same time, anecdotal evidence suggested that one native species, the Oahu 'Amakihi (*Hemignathus chloris*), appeared to be rebounding. We tested over 40 of these natives from the leeward Koolau range's Manoa Valley for malaria, and failed to find a single PCR-positive individual (Shehata *et al.* in press). Twenty of these samples were later examined by Western blots, and a single bird had evidence of a past infection based on the appearance of a protein band characteristic of birds surviving malarial challenge experiments (Atkinson, pers. comm.). These data tell us that the birds can be infected with the pathogen, but they must somehow clear the infection, and the mechanisms of resistance may be varied. Both young and adult birds are in this sample, mosquitoes were plentiful, and malarial-positive individuals from other species were found in all months that the 'Amahiki were caught and tested PCR negative. It is our hope that the population is rebounding now because a malarial-resistant or malarial-tolerant lineage of native birds has at last evolved. These birds should become the subject of challenge experiments to identify their responses to deliberate infection with malarial parasites.

Hawaiian honeycreepers represent a recent adaptive radiation that took place less than ten million years ago (Tarr and Fleischer, 1995). If genes for resistance to malaria occur in one species that is currently nonendangered, it is highly likely that homologues in some of the endangered species can be identified and mapped using standard molecular genetic techniques with microsatellites. Careful analysis of genetic lineages, to ensure that cocks and hens chosen for captive breeding include these loci, represents the only long-term strategy for conservation that will be effective. Complete removal of disease vectors is impossible, and continued importation of domestic birds guarantees that maintenance of disease-free bird stocks is unlikely.

6.5 Conclusions

Sick birds do not normally fly into mist nets, and any attempt to identify infectious disease impacts on wild bird populations will be subject to known experimental biases. This review has treated malaria as the one disease of critical importance when, in fact, avian poxvirus is transmitted by the same vector, and birds can suffer from both diseases simultaneously. New diagnostics for bird pox are being developed, and we await these methods for testing with thousands of archived DNA samples. When we have full insight into the combined effects of malaria and poxvirus, strategies for conservation could be altered. There may be changes in habitat recolonization success or lineage survival rates when birds must cope with both diseases, rather than a just a single infectious agent. Fitness costs associated with fighting both pathogens at once will no doubt alter the maintenance of genetic variation for host resistance and the evolution of parasite virulence (Yan *et al.* 1997).

In Hawaii, a combination of field and laboratory techniques targeting color-banded individuals who are bled upon capture is being used to follow the long-term reproductive success of common and rare species in various types of forest. We hope that this multidisciplinary approach will allow the discovery of new information for conservation biology about the breeding strategies and genetic characteristics of birds that would be impossible to discover from a single perspective. Infectious disease and loss of molecular diversity are only small parts of the spectrum of problems that face the remaining endangered bird populations.

However, we feel we are making progress with this approach, because we know that introduced birds from low-elevation forests are clearing malarial infections from their bloodstreams in periods as short as three months (Shehata *et al.*, in press). We know that they survive over several breeding seasons, once infected, so they are able to reproduce and expand their ranges with relative ease. We have also learned that certain core species in the bird community may be necessary for epizootic maintenance of malaria in a habitat patch when native birds are rare.

We do not as yet know the extent to which particular genotypes of pathogens can be selected for in various hosts. One recent summary of mechanisms by which avian malarial parsites are prevented from developing in their mosquito vector lists at least five different scenarios (Yan *et al.* 1997), and vertebrate hosts have equally complicated resistance strategies to malaria that require modeling quantitative traits (Wakelin 1996). Finally, studies on malarial resistance are usually done in areas where the prevalence of malarial parasites is low (Paul *et al.* 1995) compared to Hawaii. Long-term studies involving challenge experiments, hormone assays, and extensive gathering of information about the parasite genomes present are necessary to explore the immune systems of Hawaiian forest birds, the contributions of infectious disease to plumage variation and sexual selection, and the identification of those lineages most likely to survive in the changing Hawaiian ecosystems, where malaria is now a fact of life.

As tolerant or resistant host populations evolve, parasite lineages will most likely respond in step, and metapopulation analyses (Thompson 1996) for maintenance of disease resistance will assume an increased level of importance for managers. In addition, these populations provide experimental models in which to test the theoretical predictions concerning genetic characteristics of disease-resistant populations and the time for resistance loci to evolve (Dobson and May 1986). The abundant records for introductions of hosts, pathogens, and new molecular techniques to measure gene flow between populations will be our tools.

Hawaiian birds are often cited as textbook examples of the wonderful diversity of life that evolves in geographic isolation from only a few founders. As transmitters of a unique genetic heritage, these founders experienced a limited range of parasites, and their descendants faced even fewer challenges. Isolation, broken first by Polynesian colonists, then by European explorers, is no longer a positive evolutionary force for them. Native birds have now experienced over a hundred generations of life with a pathogen that kills most of them. How they are marshaling the few defenses still present and, hopefully, evolving new ones, will demonstrate to evolutionary biologists the creative aspects of extreme natural selection in changing ecosystems.

REFERENCES

Armstrong, R. W. (1973). *Atlas of Hawaii*. Honolulu: University of Hawaii Press.

Atkinson, C. T. and van Riper III, C. (1991). Pathogenicity and epizootiology of avian Haematozoa: *Plasmodium, Leucocytozoon, and Haemoproteus*. In Loye, J. E., and Zuk, M. (eds.), *Bird-Parasite Interactions*, 19–48. New York: Oxford University Press.

Atkinson C. T., Woods, K. L., Dusek, R. J., Sileo, L., and Iko, W. M. (1995). Wildlife disease and conservation in Hawaii: pathogenicity of avian malaria (*Plasmodium relictum*) in experimentally infected I'iwi (*Vestiaria coccinea*). *Parasitology, 111,* S59–S69.

Berger, A. J. (1973). *Birds*. In R. W. Armstrong (ed.), *Atlas of Hawaii*, 1st ed., 4th printing, 70–73, Honolulu: University of Hawaii Press.

Cann, R. L., Feldman, R. A., Agullana, L., and Freed, L. A. (1996). A PCR approach to detection of malaria in Hawaiian birds. In Smith, T. A. and Wayne, R. A. (eds.), *Molecular Genetic Approaches to Conservation*, 202–213. New York: Oxford University Press.

Dobson, A. P., and May, R. M. (1986). Patterns of invasions by pathogens and parasites. In Mooney, H. A., and Drake, J. A. (eds.), *Ecology of Biological Invasions of North America and Hawaii*, 58–76. New York: Springer-Verlag.

Dobson, A. P., and Miller, D. (1989). Infectious diseases and endangered species Management. *Endangered Species Update, 6(9)*, 1–5.

Ewald, P. W. (1994). *Evolution of Infectious Disease*. New York: Oxford University Press.

Feldman, R. A., Freed, L. A., and Cann, R. L. (1995). A PCR test for avian malaria in Hawaiian birds. *Molecular Ecology, 4,* 663–673.

Freed, L. A. and Cann, R. L. (1989). An integrated conservation strategy for Hawaiian forest birds. *Bioscience, 39(7),* 475–476.

Goff, L. and van Riper III, C. (1980). Distribution of mosquitoes (*Diptera: Culicidae*) on the east flank of Mauna Loa volcano, Hawaii. *Pacific Insects, 22,* 178–188.

Hamilton, W. D. (1980). Sex versus non-sex versus parasite. *Oikos 35,* 282–290.

Hayworth, A. M., van Riper III, C., and Weathers, W. W. (1987). Effects of *Plasmodium relictum* on the metabolic rate and body temperature in canaries (*Serinus canarius*). *J. of Parasitol. 73,* 850–853.

Henshaw, H. W. (1902). Birds of the Hawaiian Islands being a Complete List of the Birds of the Hawaiian Possessions with Notes on their Habits. Honolulu: Thos. G. Thrum.

Kocher, T. D., Thomas, W. K, Meyer, A., Edwards, S. V., Paabo, S., Villiblanca, F. X., and Wilson, A. C. (1989). Dynamics of mitochondrial DNA evolution in animals: amplification and sequencing with conserved primers. *Proc. Natl. Acad. Sci. USA, 86,* 6196–6200.

Kuehler, C., Kuhn, M., Kuhn, J. E., Lieberman, A., Harvey, N., and Rideout, B. (1996). Artificial incubation, hand-rearing, behavior, and release of Common 'Amakihi (*Hemignathus virens virens*): surrogate research for restoration of endangered Hawaiian forest birds. *Zoo Biology, 15,* 541–553.

Laird, M. and van Riper III, C. (1981). Questionable reports of *Plasmodium* from birds in Hawaii, with the recognition of *P. relictum spp. capistranoae* (Russell, 1932) as the avian malarial parasite there. In Canning, E. V. (ed.), *Parasitological Topics*, Soc. Cap. Protozool. Spec. Pub. *1*, 59–165. Lawrence, Kansas: Allen Press.

Loye, J. and Carroll, S. (1995). Birds, bugs and blood: avian parasitism and conservation. *Trends Ecol. Evol.*, 10(6), 232–235.

Moore, J. G., Normark, W. R., and Holcomb, R. T. (1994). Giant Hawaiian under-water landslides. *Science*, *264*, 46–47.

Moulton, M. P. and Pimm, S. L. (1983). The introduced Hawaiian avifauna: biogeo-graphic evidence for competition. *Am. Natur.*, *121(5)*, 669–690.

Paul, R. E. L., Packer, M. J., Walmsley, M., Lagog, M., Ranford-Cartwright, L. C., Paru, R., and Day, K. P. (1995). Mating patterns in malaria parasite populations of Papua New Guinea. *Science*, *269*, 1709–1711.

Perkins, R. C. L. (1903). *Fauna Hawaiiensis* or the Zoology of the Sandwich (Hawai-ian) Isles. Volume 1, Part IV, *Vertebrata*. Cambridge: Cambridge University Press.

Pratt, H. D. (1994). Avifaunal change in the Hawaiian islands, 1893–1993. In Jehl Jr., J.R., and Johnson, N.K. (eds.), *Studies in Avian Biology: A century of avigaunal change in Western North America*, 103–118, 15, Los Angeles: Cooper Ornitholog-ical Society.

Pimm, S. L. (1991). *The Balance of Nature?* Chicago: University of Chicago Press.

Price, P. W. (1980). *The Evolutionary Biology of Parasites*. Monographs in Popula-tion Biology, 15, Princeton: Princeton University Press.

Scott, J. M. (1988). The impact of infection and disease on animal populations: implications for conservation biology. *Conservation Biology*, *2(1)*, 40–56.

Scott, J. M., Mountainspring, S., Ramsey, F. L., and Kepler, C. B., eds. (1986). *Studies in Avian Biology: Forest bird communities of the Hawaiian islands: their dynamics, ecology, and conservation*. 9, Los Angeles: Cooper Ornithological Soci-ety.

Shehata, C., Freed, L. A., and Cann, R. L. (in press) Changes in native and intro-duced bird populations on Oahu: infectious disease and the emergence of resistance. In Scott, J. M., Conant, S., Freed, L. A., and van Riper III, C. (eds.), *Studies in Avian Biology*, Los Angeles: Cooper Ornithological Society.

Steadman, D. W. (1989). Extinction of birds in eastern Polynesia: a review of the record, and comparisons with other Pacific island groups. *J. of Archaeol. Sci.*, *16*, 177–205.

Tarr, C. L. and Fleischer, R. C. (1995). Evolutionary relationships of the Hawaiian honeycreepers (*Aves: Drepanidinae*). In Wagner, W. L., and Funk, V. A. (eds.), *Hawaiian Biogeography*, 147–159. Washington DC: Smithsonian.

Thompson, J. N. (1996) Evolutionary ecology and conservation of biodiversity. *Trends Ecol. Evol.*, *11(7)*, 300–303.

Thorne, E. T. and Williams, E. S. (1988). Disease and endangered species: the black-footed ferret as a recent example. *Conserv. Biol.*, *2(1)*, 66–74.

van Riper, S. G. and van Riper III, C. (1985). A summary of known parasites and diseases recorded from the avifauna of the Hawaiian islands. In Stone, C. P., and Scott, J. M. (eds.), *Hawaii's Terrestrial Ecosystems: Preservation and Manage-ment*, 298–371. Honolulu: University of Hawaii Press.

van Riper III, C., van Riper, S. G., Goff, M. L., and Laird, M. (1986). The epi-zootiology and ecological significance of malaria in Hawaiian land birds. *Ecological Monograph*, *56*, 327–344.

Wakelin, D. (1996). *Immunity to parasites, 2d ed.* Cambridge: Cambridge University Press.

Warner, R. E. (1968). The role of introduced diseases in the extinction of the endemic Hawaiian avifauna. *Condor*, *70*, 101–120.

Williams, R. (1987). Alien birds on Oahu: 1944–1985. *Elepaio*, *47*, 87–92.

Yan, G., Severson, D. W., and Christensen B. M. (1997). Costs and benefits of mosquito refractoriness to malaria parasites: implications for genetic variability of mosquitoes and genetic control of Malaria. *Evolution*, *5*, 441–450.

7

Extinction and Endangerment of Hawaiian Honeycreepers: A Comparative Approach

LEONARD A. FREED

SUMMARY. Since conservation biology is founded on general principles of ecology and evolutionary biology, it is important to use the broad methodology available to those disciplines. Here, I use the comparative method to investigate proximate and ultimate causes of extinction and endangerment of Hawaiian honeycreepers (Drepanidinae). A core community of honeycreepers existed on multiple islands and were exposed to the same threats of habitat destruction and introduced predators, competitors, and disease. Extinct species were specialized on understory plant species, with limited distribution entirely within the range of introduced disease and vectors. Endangered species are restricted now in distribution to upper elevations, where introduced disease and vectors are less prevalent. Unlisted species had a larger historical distribution, are found in more than one type of habitat, and have both larger clutch sizes and more broods per year than endangered species. The genetics underlying ecological specialization and immunity, and lower genotypic diversity associated with lower reproductive rate, are related to the ultimate causes of endangerment and extinction.

INTRODUCTION

Conservation biology is an applied endeavor grounded in the two basic disciplines of ecology and evolutionary biology (Soulé and Wilcox 1980; Frankel and Soulé 1981; Soulé and Simberloff 1986). Hutchinson (1965) clarified the relations between these disciplines in the expression "the ecological theater and the evolutionary play," in recognition of the ecological context in which evolutionary processes of selection, drift, migration, and mutation occur. Is it possible for conservation biology to be viewed as an altered evolutionary play within a damaged ecological theater? Is it possible to distinguish between proximate and ultimate causes of endangerment and extinction in the way that these types of causation are distinguished in studies of life history and behavior (Baker 1938; Tinbergen 1963; Sherman 1988)? I intend to pursue these parallels using the example of native passerine birds in the Hawaiian Islands.

The causes of extinction and endangerment of modern organisms in general are well documented. There can be direct exploitation by humans; degradation or loss of habitat; introduced predators, competitors, and disease; and fragmentation of range resulting in smaller population sizes (Diamond 1984). All identify a damaged or changed ecological theater with novel selection pressures. Often the factor(s) responsible for the decline of a particular species can be identified. But a case-by-case approach does not necessarily establish the link between the changed ecological condition and the evolutionary inability of the organism to respond adaptively to the change.

Evolutionary ecology uses the comparative method to study adaptation by identifying why some characteristics — morphological, physiological, behavioral — occur when or where they do (Darwin 1859; Brooks and McLennan 1991; Harvey and Pagel 1991). The strength of the method is based on using phenomena that are repeated independently in different locations and/or at different times by different phylogenetic lineages. The inference is that convergent evolution of morphology and behavior results from similar selection pressures. However, caution must be used to ensure that similar phenotypes are produced independently and are not simply the consequence of phylogenetic relationship (Brooks and McLennan 1991; Harvey and Pagel 1991).

Analogously, conservation biology uses comparisons to identify the general factors that contribute to extinction and endangerment. For example, islands appear to have extinction rates far greater than continents, especially when introduced predators are involved (Greenway 1967; Williamson 1981; Moors 1985). Fragmented habitats have lower diversity than continuous habitats (Wilcove *et al.* 1986; Lovejoy *et al.* 1986). Predators near the top of a food chain may be the first to decline or become extinct in relation to a disturbance (Belovsky 1987). These general statements are supported by numerous independent examples. Of course, the "convergence" in conservation biology is different than that recognized in evolutionary ecology. Taxa do not evolve characters that can be a priori identified as leading to susceptibility in the same way that taxa evolve characters that can be identified as adaptations in accordance with design criteria (Williams 1966). The major difference between convergence in conservation biology and evolutionary ecology is that a character change is involved in the latter while the decline of a taxon is involved in the former, without there necessarily being a change in character.

There are also analogies in use of proximate and ultimate causes (or factors (Thomson 1950)). A proximate cause in evolutionary ecology is a feature of the environment that elicits a physiological or behavioral response, such as a changing photoperiod which elicits initiation of breeding. A proximate cause of endangerment in conservation biology would be the changed physical or biological features of the environment that cause harm to individuals and their reproduction. An ultimate cause in evolutionary ecology is a problem that requires the appropriate performance of an organism as a solution, such as food limitation favoring timing of breeding to coincide with a flush of resources that could best support the breeding attempt. In conservation biology, an ultimate cause of endangerment or extinction would be the genetic constraints leading to inability of an organism to respond to new selective pressures generated by the proximate causes.

Disease can be used to illustrate proximate and ultimate causation and the comparative method in conservation biology. Introduced diseases and vectors could be the proximate causes of extinction, while the inability of an immune system to deal with those diseases or of behavior to reduce contact with the vector would be the ultimate cause. Whereas the comparative method in adaptation leads to predictions about performance and fitness in relation to an ultimate cause, the comparative method in conservation biology leads to a prediction about differences in characters between viable and threatened species in relation to an ultimate cause. In the context of introduced disease, the comparative method leads to a prediction that the endangered species have less genetic tolerance or resistance to pathogens than the unlisted species.

The evolution and extinction of Hawaiian honeycreepers (Drepanidinae) provide an exceptional opportunity to identify proximate and ultimate causes of extinction through the comparative method. An ecologically similar set of honeycreepers once existed in the main forest type on all of the main islands, and each main island has been exposed to roughly the same proximate factors in the form of introduced species and habitat degradation. The islands thus serve as replicates for a comparative study. In addition, because an adaptive radiation produced the honeycreepers (Freed *et al.* 1987a), the Hawaiian system is especially useful for separating the roles of phylogeny and ecological traits in making a species susceptible to extinction.

In the remainder of this chapter, I will briefly summarize features of ecology and evolution in the Hawaiian Islands, introduce the community of Hawaiian honeycreepers that co-occur in the same type of habitat found on all main islands, and identify the proximate causes of extinction and endangerment. The pattern of extinction and endangerment of birds on each island will then be analyzed with respect to a random model.

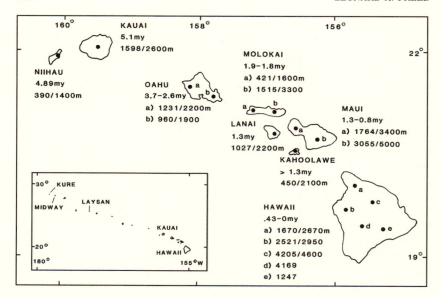

Figure 7.1. Map of the main Hawaiian Islands with inset of the entire chain. Volcanoes for each island are shown by solid circles. Accompanying information includes range of ages of volcanoes, and the highest elevation and current elevation above sea level. (Data from Carson and Clague 1995; Stearns 1985.)

Finally, the pattern will be combined with knowledge of natural history to deal with various hypotheses of ultimate causes.

7.1 PATTERNS IN THE EVOLUTION OF HAWAIIAN HONEYCREEPERS

General Background

The Hawaiian Islands are the most isolated islands in the world. The closest point to the North American continent is 3,846 kilometers. The closest point to the Asian continent is 4,900 kilometers. The entire Hawaiian chain (132 islands, reefs, and shoals) extends linearly over 2,451 kilometers, from Kure Atoll in the northwest to the newly forming island of Loihi southeast of the island of Hawaii (Fig. 7.1; Department of Geography 1983).

Formed by a hot spot over which the Pacific plate moved, the islands northwest of Loihi are progressively older (Fig. 7.1). Midway Island is approximately 15 million years old while the island of Hawaii, the youngest aerial island, is less than 500,000 years old and is still growing

(Carson and Clague 1995). The younger islands are higher than the older islands, both at present and at past maximum heights (Fig. 7.1). All of the eight main high islands now have forests and birds, except for Niihau and Kahoolawe. In addition, Molokai, Lanai, and Maui (along with Kahoolawe) form the Maui-Nui complex. During times of lower sea level, land connections existed among these particular islands (Carson and Clague 1995).

The isolation in all directions has resulted in few colonizations of organisms and rare repeats of colonization (Carlquist 1980). Subsequent divergence from source populations has resulted in extensive endemism ranging from subspecies to subfamilies of birds, bats, fishes, snails, spiders, insects, flowering plants, and ferns (Simon 1987). The biota is derived from both the Old World and the New World, guaranteeing that novel selective pressures were encountered by colonists (Wagner *et al.* 1990).

The chronology of the islands contributes to the endemism. The direction of colonizations among the Hawaiian Islands would appear to be from older islands to younger islands, although other patterns also exist (Funk and Wagner 1995; Fleischer *et al.* 1998). The distances among the main islands (excluding Niihau and Kahoolawe) range from 59 to 145 kilometers (Clague and Dalrymple 1989). Nevertheless, gene flow can be extremely restricted among populations on all islands, as evidenced by differentiation at the subspecies or species level of related forms (Wagner *et al.* 1990; Freed *et al.* 1987a). Differentiation could have resulted from founder events for individual taxa or from the order in which taxa from an older island colonized the younger. For example, a species whose morphology was shaped by competition with other species may have encountered short-term ecological release if it colonized a new island before its previous competitor. In addition, each island has some endemic genera or species that occur nowhere else.

Habitats at comparable elevations and exposure to the winds have remarkably similar biotas on the different islands (Sanderson 1993; Wagner *et al.* 1990). For example, all high main islands have (or had) mesic and wet forest with the same characteristic woody vegetation. The most widely distributed tree is *Metrosideros polymorpha* (=ohialehua [in Hawaiian]= ohia; Myrtaceae), which ranges from sea level to over 2,200 meters on islands that are that high. Ohia occurs extensively with *Acacia koa* (=koa; Fabaceae), another tree species found on all main islands. Associated with ohia and ohia/koa forest are a number of under- and midstory woody plants, including several genera of Lobelioideae (Campanulaceae) with their own adaptive radiations (Givnish *et al.* 1995; Lammers 1995) and species of *Ilex* (Aquifoliaceae),

Cheirodendron (Araliaceae), *Vaccinium* (Ericaceae), *Coprosma* (Rubi-aceae), *Myrsine* (Myrsinaceae), and *Styphelia* (Epacridaceae) that have differentiated among the main islands (Wagner *et al.* 1990). Ohia and ohia/koa forest are (or were) the most extensive habitats.

The Hawaiian honeycreepers are an endemic passerine bird subfamily (Drepanidinae) most closely related to the Carduelinae within the Fringillidae (Amadon 1950; Raikow 1976). Over sixty taxa were known historically and from the fossil record (Wilson and Evans 1890–1899; Rothschild 1893–1900; Henshaw 1902, Perkins 1903; Munro 1960; James and Olson 1991). They represent the most extreme adaptive radiation in birds in terms of plumage, bill morphology related to diverse dietary specializations, and behavior (Fig. 7.2; Freed *et al.* 1987a). Three groups were recognized historically: finch-billed birds in the tribe psittirostrini, diverse insectivorous birds in the hemignathini, and nectarivorous birds in the drepanidini (Fig. 7.2; Perkins 1903; Amadon 1950). More recently, Fleischer *et al.* (1998), using mtDNA, identified a basal split between two major clades, one consisting of the *Oreomystis* creeper from Kauai and the *Paroreomyza* creepers, the other including the psittirostrini, the drepanidini, and the remaining hemignathini. The timing of divergence, four-to-five million years, implicates Kauai as the island of origin. Thus, it is possible for the radiation producing the distinctive combinations of feeding morphology and behavior to have occurred on older islands, with subsequent colonization of younger islands as habitat became suitable. The genetic relationships of amakihis in the genus *Hemignathus* on different islands are consistent with this geological model of sequential colonization (Cann and Douglas, Chapter 6 in this volume).

Hawaiian Honeycreepers of Ohia/Koa Forest

A remarkable fact about the honeycreepers is the relative constancy of community structure among islands. I use the term "ecomorph" to indicate a particular set of morphology and behavior within a community to accommodate potential cases of convergence. Each island (Kauai, Oahu, Maui-Nui, Hawaii) had the same set of eight ecomorphs in ohia-koa forest (Table 7.1). Individual islands have or had a few endemic honeycreepers in addition to these forms in ohia/koa forest. Kauai has the anianiau (*Hemignathus parvus*) as a "lesser" amahiki. The island of Hawaii had a greater amakihi (*Hemignathus sagittirostris*), and a mamo (*Drepanis pacifica*) as a long-billed specialized nectarivore. Maui has the Maui parrotbill (*Pseudonestor xanthophrys*), an insectivore that tears and crushes wood; the crested honeycreeper (*Palmeria dolei*), a dominant nectarivore; and the Poouli (*Melamprosops phaesoma*), with creeperlike habits of feeding on snails and other invertebrates. Molokai

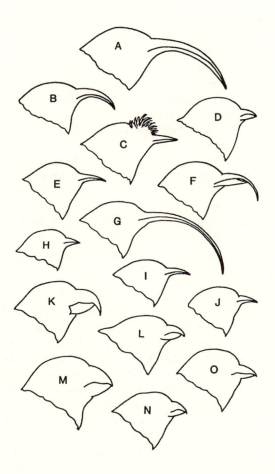

Figure 7.2. Profiles of subset of Hawaiian honeycreepers. Nectarivorous forms in the tribe Drepanidini include A, mamo (*Drepanis pacifica*); B, iiwi (*Vestiaria coccinea*); C, crested honeycreeper (*Palmeria dolei*); D, ula-ai-hawane (*Ciridops anna*); and H, apapane (*Himatione s. sanguinea*). Insectivorous forms in the tribe Hemignathini include F, akiapolaau (*Hemignathus munroi*); G, Kauai akialoa (*Hemignathus procerus*); H, Hawaii akepa (male, *Loxops c. coccinea*); I, Hawaii amakihi (*Hemignathus v. virens*); J, Kauai creeper (*Oreomystis bairdi*); and K, Maui parrotbill (*Pseudonestor xanthophrys*). Frugivorous and granivorous forms in the tribe Psittirostrini include L, ou (male, *Psittirostrra psittacea*); M, grosbeak finch (*Chloridops kona*); and N, Nihoa finch (female, *Telespyza ultima*). The poouli (O, *Melamprosops phaesoma*), recently discovered (Casey and Jacobi (1974)), has not formally been incorporated into honeycreeper phylogeny. (Outlines based on a painting by H. Douglas Pratt Jr.)

Ecomorph Name	Taxa[1]	Ecomorph Habit
Ou	*Psittirostra*	frugivore
creeper	*Paroreomyza/Oreomystis*	bark probler
Akepa	*Loxops*	leaf-bud opener
Amakihi	*Hemignathus kauaiensis/ chloris/virens*	general-foliage gleaner
Nukupuu/ akiapolaau	*Hemignathus lucidus/ munroi*	shallow excavator
Akialoa	*Hemignathus obscurus*	deep-hole prober
Apapane	*Himatione*	nectarivore, shallow flowers
Iiwi	*Vestiaria*	nectarivore, deep flowers

[1] Taxa refer to genera associated with ecomorphs. Species are indicated for *Hemignathus*, which includes several ecomorphs.

Table 7.1. Hawaiian honeycreeper ecomorphs present in ohia/koa forest on all main islands.

also had a mamo (*Drepanis funerea*). James and Olson (1991) did not attempt to locate fossil birds in ohia/koa forest because of poor fossilization conditions. However, based on their finds in dry forest and coastal situations, it is possible that some of the individual island endemics of ohia/koa forest may have been more widespread among islands.

The ecomorphs of ohia/koa forest used resources in the following way (Henshaw 1902; Perkins 1903; Munro 1960; Berger 1981). The ou consumed nectar from ohia, gleaned caterpillars from koa, and ate the fruits of understory species of lobeliods and the vine *Freycinetia arborea* (Pandaceae). These fruits are a foraging substrate unique to ou among the honeycreepers. The apapane and iiwi show the most comprehensive adaptations for nectarivory, Although the iiwi is now seen foraging mainly among ohia flowers, historical records show a strong association with lobeliods. In fact, it would be difficult to account for the length and shape of iiwi bills if they had evolved feeding on ohia flowers, which do not have corollas. A case has been made that iiwis specialized on lobeliods and other understory plants, and were prevented from feeding on ohia flowers by aggressive honeyeaters (Meliphagidae), while apapanes swamped honeyeaters by foraging in groups (Smith *et al.* 1995; Freed *et al.* 1996). Both iiwis and apapanes also consume arthropods by gleaning and occasionally flycatching (pers. observ.).

Insectivorous ecomorphs reveal extensive variation and specialization. The amakihi is the generalized gleaner, which feeds on insects and spiders on the surface of leaves and terminal twigs of ohia. Amakihis also consume nectar from ohia flowers. The creeper is the generalized gleaner of trunks and branches of ohia with its straight bill, although forms on some islands forage also as amakihis. The akepa, convergent with crossbills in the Carduelinae, with its asymmetrical jaw closure and movement, specializes in opening leaf buds of ohia to obtain microlepidopteran larvae. The nukupuu/akiapolaau feed on boring larvae, spiders, and weavils, which are found mainly in koa. Both of these species have bills specialized for excavation and use their bills (upper only in akiapolaau) to flake away lichens and loose bark to get at hidden arthropods. The akialoa, with its enormous sickle-shaped bill, fed on nectar from ohia and long flowered lobelioids and on arthropods through deep gleaning and probing in bark and holes of trees, under lichens, in stems, and in the bases of long leaves, particularly those of *Freycinetia*.

The ecomorphs vary with respect to differentiation among islands. Three taxa (ou, apapane, iiwi) are not differentiated among main islands (Amadon 1950; Tarr and Fleischer 1995). The populations of amakihi, akepa, nukupuu, and akialoa on different islands have been variably recognized at the subspecific and specific levels (Perkins 1903; Amadon 1950; Pratt 1989). Plumage, size, bill morphology, and behavior are involved in the differentiation. Particularly striking are the apparent convergences among different islands. The nukupuu, present on all islands, including Hawaii, by an overlooked single collected specimen from the expedition of Peale (1848) and a single possible recent observation (C. van Riper 1973, 1982), is similar in morphology and behavior to the akiapolaau on Hawaii except that the latter has a stouter lower bill that it uses for excavation. The creepers are remarkable in that the one species per island is from two or perhaps three different genera present statewide. Creepers of the genus *Paroreomyza* are (or were) present on Oahu, Molokai, Lanai, and Maui. Kauai has a creeper in the genus *Oreomystis*. Hawaii also has an *Oreomystis* creeper, but recent studies of mtDNA show that the Hawaii creeper is unrelated to all other creepers (Fig. 7.3, Feldman 1994). The examples of convergent evolution of ecomorphology suggest that the ecomorphs of ohia/koa forest are a core community that evolved through resource partitioning.

Geographical Evidence of the Core Community

Historical ranges of Hawaiian honeycreepers synthesized by Banko (1981–1987) provide the overlap that is required for showing that similar environments were encountered by the ecomorphs within the core community. Banko critically collated all the records of sightings and collections of Hawaiian birds between 1779 and 1979. While the islands were unevenly covered by naturalists and collectors, it is likely that the different

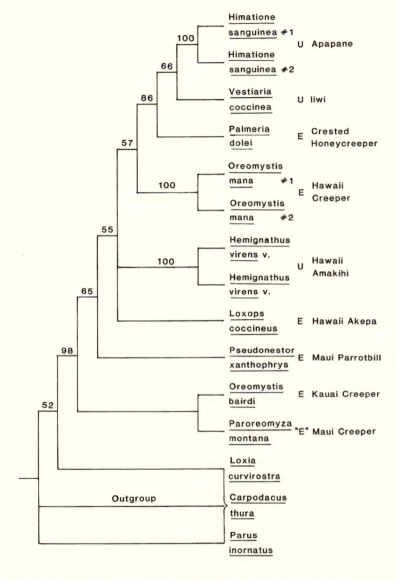

Figure 7.3. Phylogeny of subset of Hawaiian honeycreepers and outgroup. This figure is based on parsimony analyses of 790 base pairs of cytochrome *b* sequence from Feldman (1994). E = endangered, U = unlisted, "E" = unlisted but with range reduction of endangered species. Numbers associated with nodes are bootstrap values based on 1,000 replications.

KAUAI

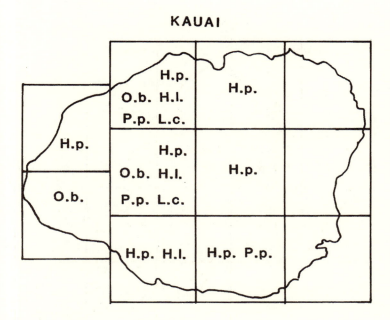

Figure 7.4. Island of Kauai with USGS quadrangles superimposed to show the historical ranges of the core endangered/extinct ohia/koa honeycreepers. Data from Banko (1981–1987). H.p. = akialoa, P.p. = ou, H.l. = nukupuu, O.b.= creeper, L.c. = akepa.

species could be detected in the locations sampled. Banko provided maps of the islands showing the distribution of sightings among United States Geological Survey (USGS) quadrangular maps.

I took the island maps and overlaid those of the core ecomorphs of honeycreepers. The result is shown in Figures 7.4-7.6 for Kauai (oldest), Oahu (intermediate), and Hawaii (youngest), respectively, for the five rarest ecomorphs (ou, akialoa, akepa, creeper, and nukupuu/akiapolaau). The Maui-Nui complex was not included because the core ecomorphs exist within the complex but not on each island, as known from historical records (but all existed on Maui when fossils are included [James and Olson 1991]). For the island of Hawaii, there were four quadrangles with all five ecomorphs, and five quadrangles with four ecomorphs. These quadrangles include the major volcanoes of Mauna Kea, Mauna Loa, and Hualalai with their extensive ohia/koa forests. For the island of Kauai, there were two quadrangles with all five ecomorphs, again in the ohia/koa forests, at the edge of the Alakai Swamp. Records are sparser

OAHU

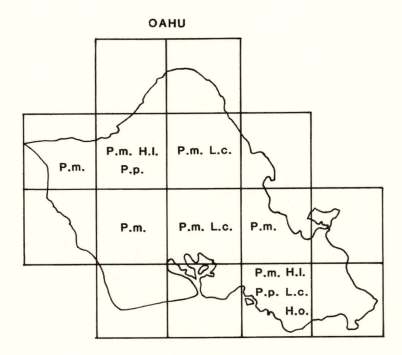

Figure 7.5. Island of Oahu with USGS quadrangles superimposed to show the historical ranges of the core endangered/extinct ohia/koa honeycreepers. Data from Banko (1981–1987). H.o. = akialoa, P.p. = ou, H.l. = nukupuu, P.m. = creeper, L.c. = akepa

for Oahu and the Maui-Nui complex, but the same pattern emerges. Ohia/koa forests have the highest overlap of ecomorphs.

A simple analysis shows that these overlapping distributions are unlikely to be due to chance. For each island — Kauai, Oahu, Hawaii — suitable habitat was defined as the superset of quadrangles in which at least one of the core ecomorphs was known historically. This excluded quadrangles that represented mountaintops above tree line or low elevations extensively cleared by humans before 1779. For each of the ecomorphs, the probability of occurring in this set of suitable habitat was defined as the proportion of the superset in which it was found. Under the null hypothesis that the geographical ranges of the ecomorphs were independent, the probability that a quadrangle in the superset contained all five would be the product of the probability for each one. This joint product represents the probability of success in a binomial distribution,

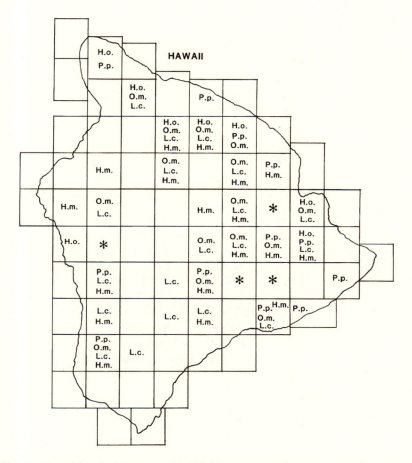

Figure 7.6. Island of Hawaii with USGS quadrangles superimposed to show the historical ranges of the core endangered/extinct ohia/koa honeycreepers Data from Banko (1981–1987). H.o. = akialoa, P.p. = ou, H.m. = akiapolaau (*Hemignathus munroi* but as *Hemignathus wilsoni* in Banko), O.m. = creeper, L.c. = akepa, * = quadrangles with all five core species.

with number of trials as the number of quadrangles in the superset, and number of successes as the number of quadrangles in the superset with all five ecomorphs.

The null hypothesis of independence is rejected for all three islands (Kauai, $p = .01$; Oahu, $p = .03$, Hawaii, $p = .04$). There are too many quadrangles with all five ecomorphs on each island. For Oahu, even one quadrangle with all five species is a significant number under the null model of independent ranges. The joint probability, under independence, is low for each island because only a small proportion of quadrangles in the superset was occupied by several of the ecomorphs. Nevertheless, this analysis shows that the concept of a core set of the five rarest ecomorphs, plus the three more common and widespread ecomorphs, is valid and that these encountered the same potential proximate threats in at least a portion of their ranges.

7.2 Extinction and Endangerment of Hawaiian Honeycreepers

Nonrandom Pattern

The decline and extinction of the core honeycreepers of ohia/koa forest on the main islands are similar. The iiwi, apapane, and amakihis are neither federally threatened nor endangered on the main islands (although the iiwi population on Oahu is listed as threatened by the state of Hawaii). The rest of the core species/subspecies/ecomorphs are extinct or endangered on all main islands, or considered for federal listing (Pratt 1994). Only the Maui creeper is considered viable in relation to the endangered species act. However, based on the extinction models for each main island in Scott *et al.* (1986), the Maui creeper shows reduction in range comparable to the endangered species with abundance comparable to the unlisted species.

This is a decidedly nonrandom pattern of unlisted and listed/extinct ecomorphs. Given that most of the honeycreepers are extinct or endangered, a model that uses the smaller class of unlisted species provides the most powerful test of nonrandomness, given the limited number of islands. The probability that apapanes, iiwis, and amakihis are unlisted on a single island is 0.375 (three of eight taxa). The probability, under a random model, that these three same taxa are unlisted on any four islands is $(0.375)^4 = 0.02$. Thus, if three of the core ecomorphs are assumed to be unlisted on each island, without specifying which ecomorphs, it is unlikely to be due to chance that the same three ecomorphs are unlisted on each island.

In addition, there is a nonrandom pattern of extinction and endangerment. The ou and akialoa are extinct on all islands; including Hawaii, Kaui, Maui, and Oahu. Under a random model, the probability of this happening is $(0.25)^4 = 0.004$. The situation is more complex with the endangered species. The akepa, creepers, and nukupuu/akiapolaau are endangered on all islands except Oahu. There, the akepa and creeper are endangered, probably extinct, and the nukupuu is extinct. Using just the three islands — Kauai, Hawaii, Maui — where they are endangered, the probability under a random model is $(.375)^3 = .05$. Therefore, the random models can be rejected. Geographical patterns of decline and extinction suggest that similar proximate and ultimate factors apply to each of the main islands.

Proximate and Ultimate Causes

Proximate factors include habitat destruction and degradation by humans and their introduced ungulates and rats, introduced mammalian and avian predators, introduced avian and vespid wasp competitors, and introduced vectors (*Culex quinquefasciatus* mosquito) and diseases (malaria [*Plasmodium relictum*] and poxvirus [*Poxvirus avium*]) (Atkinson 1977; Warner 1968; Scott *et al.* 1985; Scott *et al.* 1986). Disease is particularly relevant because different strains could have been introduced with birds from different continents, with subsequent recombination leading to virulence (see Cann and Douglas, Chapter 6 in this volume). The only major differences are that the small Indian mongoose (*Herpestes auropunctatus*) has been introduced to all islands except Kauai, and some species of introduced birds, like the white-rumped shama (*Copsychus malabaricus*), do not exist on all islands. However, to deal with ultimate factors that lead to endangerment and extinction, the viable, endangered, and extinct ecomorphs need to be compared within islands. The island of Hawaii provides the best opportunity for this, given its size, diversity of habitats, wide range of elevations, and the nature of studies conducted.

Historical distribution

The first approach to identifying ultimate factors involves comparing historical ranges. Figure 7.7 shows that the viable ecomorphs had the largest historical ranges, and extinct ecomorphs had the most limited historical ranges. Analysis of variance clearly distinguishes the three groups (mean of 14.5, 23, and 38 quadrangles, respectively, for extinct, endangered, and unlisted ecomorphs, with no overlap between groups, $p = .001$).

Another difference between the viable ecomorphs and the others, related to range, is the regular use of more than one habitat. Amakihis nest in mamane (*Sophora chrysophylla*)/naio (*Myoporum sandwicense*) forest as well as ohia-koa forest (C. van Riper 1987), and apapanes and iiwis use such forests when floral resources are abundant there (Scott *et al.* 1986). This regular use of multiple habitats is not known among the

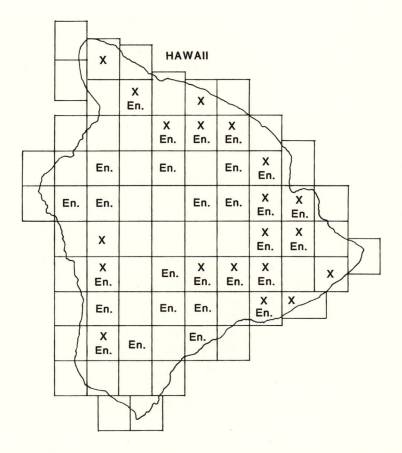

Figure 7.7. Island of Hawaii with USGS quadrangles superimposed to show historical ranges of collective endangered and extinct ohia/koa honeycreepers. Unlisted ecomorphs were present in all of these quadrangles and numerous others.

endangered or extinct ecomorphs, although the Hawaii creeper and aki-apolaau are/were extremely rare in mamane/naio forest, probably not as breeders (Scott *et al.* 1986; Snetsinger 1995).

Smaller historical range, and limitation to a single habitat, have several implications for extinction and endangerment. First, there are fewer refugia from novel physical or biological threats within the environment. Second, there is less isolation by distance that could maintain genetic heterogeneity (Wright 1943). This in turn reduces the likelihood of some individuals that might be preadapted to cope with the changed environment. In addition, existence limited to a single habitat likely leads to increased specialization that precludes adapting to changing conditions. For example, endangered ecomorphs have longer fledgling periods (time from leaving the nest to termination of parental care) than the unlisted taxa (less than one month for unlisted ecomorphs; longer for endangered ecomorphs up to twenty weeks [pers. obs.]). The longer fledgling periods of creeper, akepa, and akiapolaau, in that order, are associated with increasing morphological and behavioral specializations for extracting hidden insects (Fig. 7.2).

Changes from historical distribution

The second approach to identifying ultimate factors involves comparing the changes from historical to recent distribution. All species had more restricted recent distributions based on surveys conducted around the turn of the twentieth century (Wilson and Evans 1890–1899, Rothschild 1893–1900, Henshaw 1902; Perkins 1903), and further restrictions based on the comprehensive Hawaiian Forest Bird Survey conducted during the late 1970s and early 1980s (Scott et al. 1986). The birds now considered extinct (ou, akialoa) were extremely rare, and the birds now considered endangered or proposed for listing are limited in distribution to upper elevations. The unlisted birds are rarer at lower elevations where they formerly were common.

Scott et al. (1986) proposed a disease model emphasizing mosquito-transmitted diseases (malaria, poxvirus) to account for the change in distribution and abundance of all ecomorphs. Hawaiian honeycreepers have been shown to be more susceptible to malaria than continental birds introduced to Hawaii (C. van Riper et al. 1986). Unfortunately, experimental studies have been done only on the viable set of ecomorphs from ohia/koa forests. However, the distribution of the endangered ecomorphs is consistent with the mosquito/disease model (distribution primarily above the 1,500 meter mosquito zone [Scott et al. 1986]), many specimens of the endangered and extinct ecomorphs showed signs of active pox lesions when collected (Perkins 1903), and the viable birds are most abundant where they coexist with endangered ecomorphs at upper elevations (Scott et al. 1986). Cann and Douglas (Chapter 6 in this volume) develop the disease model more fully.

The ultimate factor pertaining to disease concerns the immune system of Hawaiian honeycreepers. The assumption has been that these birds have evolved in isolation from the majority of pathogens and vectors encountered by continental birds. The isolation would have resulted in relaxed selection that would otherwise shape an immune system. Under this model, the documented susceptibility to disease is the ultimate factor for endangerment associated with limited distribution at upper elevations or extinction at lower elevations. The low amount of Major Histocompatibility Complex (MHC) diversity in iiwi reported by Jarvi *et al.* (1999) is consistent with the susceptibility. However, the sequence of 250 base pairs of the cytochrome b gene in fifteen iiwis produced only one haplotype, in contrast to five other honeycreepers with at least four individuals sampled (Feldman 1994). Caution is necessary to distinguish the genetic effects of relaxed selective pressures from those shaped by prehistoric population bottlenecks.

Distinguishing alternative causes

Analysis of alternative proximate and ultimate causes of extinction and endangerment can now be performed. The extinct ou and akialoa specialized in feeding on the vine *Freycinetia*, a plant that was itself restricted to lower and mid elevations within ohia/koa forests, and on lobeliods. The latter became extremely rare from herbivory by feral ungulates and rats. Specialization on a food resource with limited distribution or vulnerability to novel herbivores is at least one ultimate factor responsible for extinction (see Pimm and Pimm (1982) for an example with honeyeaters in Hawaii). Rarity of that food resource from anthropogenic or even geological causes (i.e., active lava flows) could then be the proximate factor. However, the birds' ranges were well within the range of the introduced mosquito vector of introduced avian malaria and poxvirus (C. van Riper *et al.* 1986). The behavior of using introduced plants associated with coastal populations of humans (Perkins 1903) also contributed to exposure to disease as another proximate factor. Susceptibility to that disease would be another ultimate factor. If the birds died of disease without being limited by food, then an interesting interaction results. The ultimate factor of susceptibility to disease may have resulted in extinction, but the ultimate factor of specializing on a food resource that happened to be in an area with introduced disease would have prevented any escape from exposure to disease.

The endangered akepas, creepers, and nukupuus/akipolaau also have an interaction between specialization and disease. Loss of range at lower elevations is attributed to susceptibility to disease as an ultimate factor. These birds are also specialized on arthropods in ohia/koa forest, and are found in highest density in old-growth forest at upper elevations. The specialization on large trees in old-growth forest could be an ultimate factor of endangerment if such trees are not in sufficient abundance to support viable populations of the birds, or are not regenerating due to anthropogenic disturbance such as cattle ranching. The interaction between specialization and disease may be reversed from that identified for the extinct ecomorphs. For the endangered birds, disease as an ultimate factor may have restricted distribution, with consequences of limited habitat for specialized resources leading to endangerment and perhaps extinction. For the extinct ecomorphs, disease may have caused the death of individuals.

Role of Life History

Since ultimate factors of decline have implications about selective pressures and the potential to respond to them, it is relevant to compare life-history characteristics of the ecomorphs in the core community. Other things being equal, a greater diversity of genotypes would be associated with a higher reproductive rate in sexually reproducing species (Williams 1975), and there would be increased numbers of zygotes with mutations each generation. Life-history characters, such as clutch size and the number of clutches per year, can thus influence the ability of populations to respond to selective pressures (Fisher 1958). Unfortunately, nothing is known of these life-history characters in the extinct ou and akialoa. The only relevant comparisons are those between the endangered and unlisted species. Nevertheless, there are striking differences between the unlisted and endangered birds in terms of clutch size and number of broods per year.

The reproductive rate of the endangered birds is lower than that of the unlisted species. Clutch sizes of the endangered ecomorphs are smaller on average, in mode, and in maximum from unlisted birds in ohia/koa forests on Kauai (Table 7.2). Endangered birds in ohia/koa forest on the island of Hawaii also rarely lay more than two eggs (Sincock and Scott 1980; Sakai and Johanos 1983; Freed et al. 1987b; Banko and Williams 1993; Lepson and Freed 1997).

In addition, the endangered taxa of the core ecomorphs have only one brood per year, whereas the unlisted taxa have several broods per year. Mist-netting operations in ohia/koa forest on the island of Hawaii, where all extant ecomorphs exist, show lower proportions of endangered birds that are juveniles or fledglings during the months of June and July (mean 0.33 for endangered; 0.73 for unlisted; $p = .001$). This difference is associated with breeding seasons that usually extend over a three-month period (April through June) compared to a six-month period (December through June). The longer fledgling periods of the endangered ecomorphs may obviate a second brood. The single clutch may be related to the lengthy period of parental care associated with learning foraging skills (Ashmole and Humberto 1968)

Species	Sample Size	Mode	Mean, Std.	Maximum
Unlisted				
Amakihi	21	3	2.95 +/− .65	4
Apapane	38	3	2.74 +/− .41	4
Iiwi	15	2	2.00 +/− .29	3
Endangered				
Creeper	2	1,2	1.50	2
Akepa	1	2	2.0	2

[1] Data from Eddinger 1970, 1972a,b.

Table 7.2. Comparison of clutch size of honeycreepers of ohia/koa forest on Kauai

With lower reproductive rate, the endangered taxa, and perhaps the extinct species, would have fewer opportunities for natural selection to overcome drift in response to changes in the environment. This would certainly constrain them to the specializations acquired during the adaptive radiation. The key point here is that the genetic underpinnings of life-history characteristics associated with foraging and habitat specializations appear to constrain the ability to respond to environmental changes. The same constraint based on genetic underpinnings of the immune system could pertain to evolution of tolerance or resistance to introduced diseases. It is thus relevant that van Riper *et al.* (1986) found evidence of tolerance to malaria among the amakihis, iiwis, and apapanes that were experimentally challenged, and even resistance in one population of Hawaii amakihi (*Hemignathus virens virens*). Cann and Douglas (Chapter 6 in this volume) document likely resistance in the Oahu amakihi (*Hemignathus chloris*). Unfortunately, no comparable studies have been done on creepers, akepas, and akiapolaaus, which are now restricted to elevations where diseases are less of a selective pressure.

Role of Phylogeny

Formal cladistic analysis of Hawaiian honeycreepers includes mitochondrial DNA sequences and fragments of most extant taxa (Feldman 1994; Tarr and Fleischer 1995; Fleischer *et al.* 1998). While most shallow nodes remain unresolved, there are intriguing possibilities that phylogeny may be involved. Figure 7.3 shows a phylogeny based on parsimony analyses of cytochrome b sequence (Feldman 1994). Endangered status is shown for each taxon. The bootstrap value of 55 that forms a node with akepa, amakihi, creeper, and nectarivorous birds (iiwi and apapane) is weak support of the tree as drawn, showing the nectarivorous birds as the most genetically derived forms, with the amakihi in a more basal position. Consistent with the Neighbor Joining tree in Feldman (1994), the amakihis, Hawaii creeper, and the nectarivorous birds could branch off in any order. It is thus possible that the unlisted species of the core group of honeycreepers form most of a clade near the tip of the tree. This would be remarkable, indicating that the most recently evolved birds are the most general in terms of habitats used and have the highest reproductive rates among the core group of honeycreepers of ohia/koa forest.

Other aspects of endangerment suggest an alternative to phylogeny. For example, the Maui creeper (*Paroreomyza*), which is not considered endangered but shows the reduction in range of the endangered birds on Maui, is phylogenetically basal to both unlisted and endangered core honeycreepers (Fig. 7.3). This creeper differs from the others in that it gleans arthropods from foliage (like amakihis), as well as bark and branches. It may even have a higher realized reproductive rate than creepers on other islands, given its abundance, so both of these characteristics, shared with other unlisted species, may pertain to its status. Another example, the endangered crested honeycreeper (*Palmeria dolei*), while not a core honeycreeper, is nevertheless closely related to the unlisted core honeycreepers (Fig. 7.3). This indicates that endangerment can be recognized in the recently evolved clade. The crested honeycreeper also is the one known exception in ohia/koa forest of an endangered species with multiple broods per year (VanGelder and Smith 1999).

7.3 CONCLUSIONS

Hawaiian honeycreepers of ohia/koa forest represent perhaps the clearest system to which the comparative approach in conservation biology can be applied. Other insular systems, where the same taxa are present on several islands, could be used, even if the chronological sequence of the islands or colonizations is not known. Continental systems can also be used, because the "islands" are being created by fragmentation (Temple 1986). This sets up the replicates with similar habitat and ecomorphs for analysis.

Consideration of proximate and ultimate causes of extinction and endangerment can have implications for conservation management. Most actions that have been proposed for Hawaiian birds, and for organisms elsewhere in the world, involve the proximate causes. This action includes removing predators, preserving habitat, restoring damaged habitat, and controlling vectors of disease. However, addressing ultimate causes may be more successful in the long run, drawing on principles associated with evolution in changing environments (Levins 1968; Lomnicki 1988).

The tactic of double-clutching, wherein the first clutch of eggs is taken for rearing in captivity and the replacement clutch is tended by the parents, has the effect of increasing genotypic diversity as well as number of individuals through higher reproductive rate. Perhaps some of this diversity could result in better use of habitat in the current condition as proximate factors are managed. In addition, to the extent that specialization is learned from parents, the captive breeding operation has the opportunity to influence, adaptively for current conditions, the learning of the subjects.

Another direct approach to managing ultimate factors involves disease. Natural variation in susceptibility to disease is known to exist in most populations of Hawaiian honeycreepers that have been examined (van Riper *et al.* 1986). As Cann and Douglas (Chapter 6 in this volume) emphasize, individual birds could be tested for tolerance and resistance to malaria or poxvirus. The tolerant or resistant birds could then be used as the sires and dams in captive breeding programs. The offspring of these birds might then be used to establish new populations within the historical range or to augment gene pools that are deficient in the lineages of those birds.

ACKNOWLEDGMENTS

Many thanks to Laura Landweber and Andrew Dobson for the invitation to participate in this symposium, which also provided an opportunity to get together once again with Steve Hubbell and Leslie Johnson, my thesis advisors. I dedicate this paper to Leslie's memory, and wish I had the opportunity to discuss it with her. Thanks also to Jaan Lepson, who helped collect much of the life-history data; Bob Feldman for permission to use Figure 7.3 from his dissertation; Doug Pratt for permission to use outlines of his picture of honeycreepers; Sue Monden for figures; Becky Cann for helping to clarify some of the issues; Rae Miller for helpful comments; and several anonymous reviewers.

References

Amadon, D. (1950). The Hawaiian honeycreepers (Aves, Drepaniidae). *Bull. Am. Mus. Nat. Hist., 95*, 151–262.

Ashmole, N. P., and Humberto, T. S. (1968). Prolonged parental care in royal terns and other birds. *Auk, 85*, 90–100.

Atkinson, I. A. E. (1977). A reassessment of the factors, particularly *Rattus rattus* L., that influenced the decline of endemic forest birds in the Hawaiian Islands. *Pac. Sci., 31*, 109–133.

Banko, P. C., and Williams, J. (1993). Eggs, nest, and nesting behavior of akiapolaau (Drepanidinae). *Wilson Bull., 105*, 427–435.

Banko, W. E. (1981–1987). *History of Endemic Hawaiian Birds. Part 1. Population histories — species accounts.* Avian History Reports 8–11. Honolulu: Cooperative National Park Resources Studies Unit, University of Hawaii.

Baker, J. R. (1938). The evolution of breeding seasons. In de Beer, C. (ed.), *Evolution: Essays on aspects of evolutionary biology,* 161–177, Oxford: Oxford University Press.

Belovsky, G. E. (1987). Extinction models and mammalian persistence. In Soulé, M. (ed.), *Viable Populations for Conservation,* 35–57, Cambridge: Cambridge University Press.

Berger, A. J. (1981). *Hawaiian Birdlife, 2d ed.* Honolulu: University of Hawaii Press.

Brooks, D. R., and McLennan, D. A. (1991). *Phylogeny, Ecology, and Behavior.* Chicago: University of Chicago Press.

Carlquist, S. (1980). *Hawaii, A Natural History: Geology, climate, native flora and fauna above the shoreline, 2d ed.* Lawai, Kauai: Pacific Tropical Botanical Garden.

Carson, H. L., and Clague, D. A. (1995). Geology and biogeography of the Hawaiian Islands. In Wagner, W. L., and V. A. Funk, V. A. (eds.), *Hawaiian Biogeography: Evolution on a hot spot archipelago,* 14–29, Washington, D.C.: Smithsonian Institution Press.

Casey, T. L. C., and Jacobi, J. D. (1974). A new genus and species of bird from the Island of Maui, Hawaii (Passeriformes: Drepanididae). *Occas. Pap. Bernice P. Bishop Mus., 24*, 216–226.

Clague, D. A., and Dalrymple, G. B. (1989). Tectonics, geochronology, and origin of the Hawaiian-Emperor Volcanic Chain. In Winterer, E. L., Hussong, D. M., and Decker, R. W. (eds.), *The eastern Pacific Ocean and Hawaii. The Geology of North America,* N:188–217, Boulder, CO: The Geological Society of America.

Darwin, C. (1859). *On the Origin of Species by Means of Natural Selection, or, the Preservation of Favoured Races in the Struggle for Life.* London: John Murray.

Department of Geography. (1983). *Atlas of Hawaii, 2d ed.* Honolulu: University of Hawaii Press.

Diamond, J. A. (1984). Historic extinctions: a Rosetta Stone for understanding prehistoric extinctions. In Martin, P. S., and Klein, R. G. (eds.), *Quaternary Extinctions: A prehistoric revolution,* 824–862, Tucson: University of Arizona Press.

Eddinger, C. R. (1970). A study of the breeding biology of four species of Hawaiian honeycreepers (Drepanididae). Ph. D. dissertation, University of Hawaii.

Eddinger, C. R. (1972a). Discovery of the nest of the Kauai akepa. *Wilson Bull., 84*, 95–97.

Eddinger, C. R. (1972b). Discovery of the nest of the Kauai creeper. *Auk, 89*, 673–674.

Feldman, R. A. (1994). Molecular evolution, genetic diversity, and avian malaria in the Hawaiian honeycreepers. Ph. D. dissertation, University of Hawaii.

Fisher, R. A. (1958). *The Genetical Theory of Natural Selection, 2d ed.* New York: Dover.

Fleischer, R. C., McIntosh, C. E., and Tarr, C. L. (1998). Evolution on a volcanic conveyor belt: using phylogeographic reconstructions and K-Ar-based ages of the Hawaiian Islands to estimate molecular evolutionary rates. *Mol. Ecol., 7,* 533–545.

Frankel, O. H., and Soulé, M. E. (1981). *Conservation and Evolution.* Cambridge: Cambridge University Press.

Freed, L. A., Conant, S., and Fleischer, R. C. (1987a). Evolutionary ecology and radiation of Hawaiian passerine birds. *Trends Ecol. Evol., 2,* 196–203.

Freed, L. A., Telecky, T. M., Tyler III, W. A., and Kjargaard, M. A. (1987b). Nest-site variability in the akepa and other cavity-nesting birds on the island of Hawaii. *Elepaio, 47,* 79–81.

Freed, L. A., Smith, T. B., Carothers, J. H., and Lepson, J. K. (1996). Shrinkage is not the most likely cause of bill change in iiwi: a rejoinder to Winker. *Conserv. Biol., 10,* 659–660.

Funk, V. A., and Wagner, W. L. (1995). Biogeographic patterns in the Hawaiian islands. In Wagner, W. L., and Funk, V. A. (eds.), *Hawaiian Biogeogrphy: Evolution on a hot spot archipelago,* 379–419, Washington, D.C.: Smithsonian.

Givnish, T. J., Sytsma, K. J., Smith, J. F., and Hahn, W. J. (1995). Molecular evolution, adaptive radiation, and geographic speciation in *Cyanea* (Campanulaceae, Lobelioideae). In Wagner, W. L., and Funk, V. A. (eds.), *Hawaiian Biogeography: Evolution on a hot spot archipelago,* 288–337. Washington, D.C.: Smithsonian.

Greenway, J. C., Jr. (1967). *Extinct and Vanishing Birds of the World.* New York: Dover.

Harvey, P. H., and Pagel, M. D. (1991). *The Comparative Method in Evolutionary Biology.* New York: Oxford University Press.

Henshaw, H. W. (1902). *Birds of the Hawaiian Islands Being a Complete List of the Birds of the Hawaiian Possessions with Notes on their Habits.* Honolulu: Thomas G. Thrum.

Hutchinson, G. E. (1965). *The Ecological Theater and the Evolutionary Play.* New Haven, CT: Yale University Press.

James, H. F., and Olson, S. L. (1991). Descriptions of thirty-two new species of birds from the Hawaiian Islands: part II. Passeriformes. *Ornith. Monogr., 46.*

Jarvi, S. I., Atkinson, C. T., and Fleischer, R. C. (1999). Immunogenetics and resistance to avian malaria (*Plasmodium relictum*) in Hawaiian honeycreepers (Drepanidinae). In Scott, J. M., van Riper, C., Freed, L., Conant, S. (eds.), *Studies in Avian Biology: Historical Ecology and Conservation of the Hawaiian Avifauna,* in press, Los Angeles: Cooper Ornithological Society.

Lammers, T. G. (1995). Patterns of speciation in *Clermontia* (Campanulaceae, Lobeliodeae). In Wagner, W. L., and Funk, V. A. (eds.), *Hawaiian Biogeography: Evolution on a hot spot archipelago,* 338–362. Washington, D.C.: Smithsonian.

Lepson, J. K., and Freed, L. A. (1997). Akepa (*Loxops coccineus*). In Poole, A. and Gill, F. (eds.), *The Birds of North America, No. 294.* Philadelphia: The Academy of Natural Sciences, and Washington, D.C.: American Ornithologists' Union.

Levins, R. (1968). *Evolution in Changing Environments.* Princeton: Princeton University Press.

Lomnicki, A. (1988). *Population Ecology of Individuals.* Princeton: Princeton University Press,

Lovejoy, T. E., Bierregaard Jr., R. O., Rylands, A. B., Malcolm, J. R., Quintela, C. E., Harper, L. H., Brown Jr,, K. S., Powell, A. H., Powell, G. N. V., Schubart, H. O. R., and Hays, M. B. (1986). Edge and other effects of isolation on Amazon forest fragments. In Soulé, M. (ed.), *Conservation Biology*, 257–285, Sunderland, Mass.: Sinauer.

Moors, P. J., Ed. (1985). *Conservation of Island Birds*. International Council for Bird Preservation Technical Publication No. 3. Norwich, UK: Page Brothers Ltd.

Munro, G. C. (1960). *Birds of Hawaii, 2d Edition*. Rutland, Vt. and Tokyo, Japan: Charles E. Tuttle Co., First published in 1944.

Peale, T. R. (1848). *United States exploring expedition, 1838–42. Mammalia and Ornithology*, 8, Philadelphia: C. Sherman.

Perkins, R. C. L. (1903). *Fauna Hawaiiensis or the Zoology of the Sandwich (Hawaiian) Isles. Vol. 1, Pt.IV, Vertebrata*. Cambridge: Cambridge University Press.

Pimm, S. L., and Pimm, J. W. (1982). Resource use, competition, and resource availability in Hawaiian honeycreepers. *Ecology*, *63*, 1468–1480.

Pratt, H. D. (1989). Species limits in akepas (Drepanidinae: *Loxops*). *Condor*, *91*, 933–940.

Pratt, H. D. (1994). Avifaunal change in the Hawaiian Islands, 1893–1993. *Studies in Avian Biology No. 15*, 103–118.

Raikow, R. J. (1976). The origin and evolution of the Hawaiian honeycreepers (Drepanidiidae). *Living Bird*, *15*, 95–117.

Rothschild, W. (1893–1900). *The Avifauna of Laysan and Neighboring Islands with a Complete History to Date of the Birds of the Hawaiian Possessions*. London: R. H. Porter.

Sakai, H. F., and Johanos, T. C. (1983). The nest, eggs, young, and aspects of the life history of the endangered Hawaii creeper. *West. Birds*, *14*, 73–84.

Sanderson, M., Ed. (1993). *Prevailing Trade Winds: Climate and weather in Hawaii*. Honolulu: University of Hawaii Press.

Scott. J. M., Kepler, C. B., and Sincock, J. L. (1985). Distribution and abundance of Hawai'i's endemic land birds: conservation and management strategies. In Stone, C. P., and Scott, J. M. (eds.), *Hawai'i's Terrestrial Ecosystems: Preservation and management*, 75–104. Cooperative National Park Resources Study Unit, Honolulu: University of Hawaii.

Scott, J. M., Mountainspring, S., Ramsey, F. L., and Kepler, C. B. (1986). *Forest bird Communities of the Hawaiian Islands: Their dynamics, ecology, and conservation*. Studies in Avian Biology, 9, Cooper Ornithological Society.

Sherman, P. W. (1988). The levels of analysis. *Anim. Behav.*, *36*, 616–618.

Simon, C. (1987). Hawaiian evolutionary biology: an introduction. *Trends Ecol. Evol.*, *2*, 175–178.

Sincock, J. L., and Scott, J. M. (1980). Cavity nesting of the akepa on the island of Hawaii. *Wilson Bull.*, *92*, 261–263.

Smith, T. B., Freed, L. A., Lepson, J. K., and Carothers, J. H. (1995). Evolutionary consequences of extinctions in populations of a Hawaiian honeycreeper. *Conserv. Biol.*, *9*, 107–113.

Snetsinger, T. J., (1995). Observations of a Hawai'i creeper in mamane forest. *Elepaio*, *55*, 55-56.

Soulé, M. E., and Simberloff, D. (1986). What do genetics and ecology tell us about the design of nature reserves? *Biol. Cons.*, *35*, 19–40.

Soulé, M. E., and Wilcox, B. A. (1980). *Conservation Biology: An evolutionary-ecological approach*. Sunderland, Mass.: Sinauer.

Stearns, H. T. (1985). *Geology of the State of Hawaii.* 2d ed. Palo Alto, CA: Pacific Books.

Tarr, C. L., and Fleischer, R. C. (1995). Evolutionary relationships of the Hawaiian honeycreepers (Aves, Drepanidinae). In Wagner, W. L. and Funk, V. A. (eds.), *Hawaiian Biogeogrphy: Evolution on a hot spot archipelago*, 147–159. Washington, D.C.: Smithsonian.

Temple, S. A. (1986). Predicting impacts of habitat fragmentation on forest birds: A comparison of two methods. In Verner, J., Morrison, M. L. and Ralph, C. J. (eds.), *Wildlife 2000: Modelling habitat relatationships of terrestrial vertebrates*, 301–304. Madison: University of Wisconsin Press.

Thomson, A. L. (1950). Factors determining the breeding seasons of birds: an introductory review. *Ibis, 92*, 17–184.

Tinbergen, N. (1963). On aims and methods of ethology. *Zeits. Tierpsych., 20*, 410–433.

VanGelder, E. M., and Smith, T. B. (1999). Breeding characteristics of the akohekohe on east Maui. In Scott, J. M., van Riper III, C., Freed, L. A. and Conant, S.(eds.), *Studies in Avian Biology: Historical Ecology and Conservation of the Hawaiian Avifauna*, in press, Los Angeles: Cooper Ornithological Society.

van Riper III, C. (1973). Island of Hawaii land bird distribution and abundance. *Elepaio, 34*, 1–3.

van Riper III, C. (1982). Censuses and breeding observations of the birds on Kohala Mountain, Hawaii. *Wilson Bull., 94*, 463–476.

van Riper III, C. (1987). Breeding ecology of the Hawaii common amakihi. *Condor, 89*, 85–102.

van Riper III, C., van Riper, S. G., Goff, M. L., and Laird, M. (1986). The epizootiology and ecological significance of malaria in Hawaiian land birds. *Ecol. Monogr., 56*, 327–344.

van Riper, S. G., and van Riper III, C. (1985). A summary of known parasites and diseases recorded from the avifauna of the Hawaiian Islands. In Stone, C. P. and Scott, J. M. (eds.), *Hawai'i's Terrestrial Ecosystems: Preservation and management*, 298–371. Honolulu: Cooperative National Park Resources Studies Unit, University of Hawaii.

Wagner, W. L., Herbst, D. R., and Sohmer, S. H. (1990). *Manual of the Flowering Plants of Hawaii, Vols 1–2.* Honolulu: University of Hawaii Press.

Warner, R. E. (1968). The role of introduced diseases in the extinction of the endemic Hawaiian avifauna. *Condor, 70*, 101–120.

Wilcove, D. S., McLellen, C. H., and Dobson, A. P. (1986). Habitat fragmentation in the temperate zone. In Soulé, M. (ed.), *Conservation Biology*, 237–256. Sunderland, Mass.: Sinauer.

Williams, G. C. (1966). *Adaptation and Natural Selection.* Princeton: Princeton University Press.

Williams, G. C. (1975). *Sex and Evolution.* Princeton: Princeton University Press.

Williamson, M. (1981). *Island Populations.* Oxford: Oxford University Press.

Wilson, S. B., and Evans, A. H. (1890–1899). *Aves Hawaiiensis: The birds of the Sandwich Islands.* London: R. H. Porter.

Wright, S. (1943). Isolation by distance. *Genetics, 28*, 114–138.

8

Something Old for Something New: The Future of Ancient DNA in Conservation Biology

LAURA F. LANDWEBER

SUMMARY. While all other chapters in this volume discuss the preservation of endangered species, the final chapter considers primarily species that are already extinct. But as this chapter will endeavor to demonstrate, some of the techniques involved in the retrieval of DNA from archaelogical remains, zoological collections, and paleontological remains may have a strong future in conservation genetics.

This chapter will focus first on some of the technical issues involved in amplifying DNA that is damaged and modified in different ways. We then describe several experiments that have used ancient DNA, marching progressively backwards in time, from present-day endangered species to ancient remains where the technical problems become increasingly more challenging. We finally conclude on an optimistic note introducing the chemical tools that provide a standard method for the identification of remains that are most likely to contain DNA.

INTRODUCTION TO ANCIENT DNA

The first step towards retrieving DNA from ancient tissues is often examination of the tissue under a light microscope. For example, a human muscle that has been air dried and rehydrated in some ways imitates mummification in ancient Egypt. However a modern muscle will display much more detail, with visible cell nuclei and striation, whereas a muscle from a two to three-thousand-year-old human will typically display hardly any intracellar structure at all.

Fortunately though, some ancient remains are better preserved. Pääbo (1985) was the first to discover DNA in an Egyptian mummy from 2400 B.C.E. Under the light microscope a skin sample revealed structures that resembled cell nuclei and were stained brightly with ethidium bromide (a dye that intercalates with DNA); (Fig. 8.1). Only 2 in 110 mummies

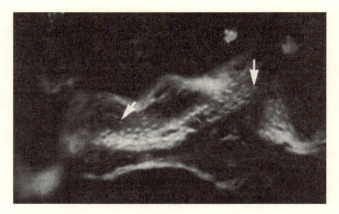

Figure 8.1. Tissue section of a skin sample stained with Ethidium Bromide. Arrows indicate bright staining of nucleic acids in the cell nuclei (reproduced with permission from *Nature*, Pääbo 1985. Copyright 1985 Macmillan Magazines Limited).

examined yielded DNA (one in 23 in the initial survey); however, when DNA was preserved, Pääbo recovered approximately 20 micrograms (20 μg) of DNA per gram of dried tissue, or about 1/20th the typical recovery from fresh tissues.

DNA in mummified tissue is usually degraded to a very small size — one hundred to two hundred base pairs — whereas DNA from modern extracts is typically tens of thousands of base pairs. However, fragments as long as 3.4 kilobases have occasionally been recovered from ancient DNA (Pääbo 1985). A variety of groups have sequenced loci of particular interest to conservation geneticists — microsatellite repeats and HLA (histocompatibility) genes from mummified archaeological remains as much as 7,500 years old (Roewer *et al.* 1991; Lawlor *et al.* 1991).

8.1 MOLECULAR ARCHAEOLOGY

The reconstruction of ancient sequences is a laborious process. Unlike modern DNA, from which one can often amplify fragments of several thousand base pairs, the study of ancient DNA requires puzzling together several short overlapping fragments. Mitochondrial DNA is one of the most commonly amplified targets because of its high copy number. This increases the likelihood of recovering a complete sequence from an overlapping set of surviving DNA molecules. Single copy genes almost always fail to produce reliable results.

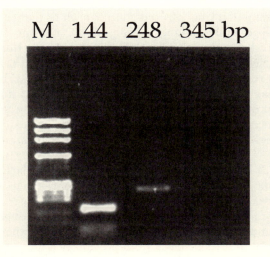

Figure 8.2. Amplification of a fragment of the mitochondrial 16S rRNA gene. PCR primers are separated by increasing distance. This demonstrates an inverse relationship between fragment length and amplification yield; there is almost no visible product at 345 bp. M is a marker lane. DNA was extracted from the bone of a giant ground sloth (adapted from Handt *et al.* 1994a).

The trend one should observe when amplifying ancient DNA is increased efficiency of amplification with shorter fragments. Thus, a 144 base pair (bp) fragment will amplify better than a 248 bp fragment, which will amplify better than a 345 bp fragment, respectively, and so on (Fig. 8.2). Amplifications from reliable sources of ancient DNA generally get progressively weaker as the distance increases between the primers. On the other hand, for modern DNA, longer amplifications often produce greater yields than shorter fragments. When this is observed for a putative "ancient" sample, it is most assuredly a sign of contamination. For example, amplification of nuclear 18S rRNA sequences with universal rRNA primers can detect fungal or bacterial contamination of ancient material (Höss *et al.* 1996a). Of course, the identity of a sequence belonging to the amplified fragment is unknown until its sequence is compared against the existing database. Therefore robust phylogenetic analysis of the sequences is essential for revealing any source of contamination.

The best example of contamination was the reported claim of dinosaur DNA, which will be discussed at the end of this chapter. A common culprit is contamination with modern sources of DNA. Of course, one can reduce contamination from obvious sources such as human DNA by using primers specifically designed to amplify DNA from the taxonomic group of interest. On the other hand, Polymerase Chain Reaction (PCR) primers designed to amplify only human sequences can detect contamination from researchers or museum curators who may have handled the specimens. Unfortunately, it is often the case that human-specific primers will produce a product but that the correctly targeted (species-specific) primers will produce only artifacts; this indicates the lack of sufficiently preserved ancient DNA in an otherwise contaminated sample.

In work with human remains, this problem is particularly severe, as it is especially hard to exclude the possibility of contamination. Logically, one takes several precautions to avoid contaminating the material and to perform every possible control. It is sometimes even advisable to work in separate laboratories to avoid contamination with either present-day related material or ancient DNA from previously studied specimens.

One can apply several criteria to test for contamination (Austin *et al.* 1997). Negative control amplifications are essential, testing every possible component as well as extracts from surfaces and material that should not contain DNA. Positive controls include the preparation of multiple DNA extracts in different laboratories. Indeed, the strongest proof of authenticity is when several groups achieve the same results in independent laboratories. The most common source of contamination is generally contamination of reagents from other experiments in the laboratory. However, it is virtually impossible to eliminate contamination of samples that are received from other laboratories. Even more insidious is a carrier effect such that low levels of contamination do not appear in negative controls because the few contaminating molecules stick to the walls of the PCR tube. However, the actual combination of an extract containing damaged ancient DNA with a contaminating source of modern DNA allows the damaged DNA molecules to coat the walls of the PCR tube. This enables detection of the rare contaminating DNA molecules in solution, which preferentially amplify over damaged DNA. In this particular situation the negative controls look clean, but the amplified band still derives from contamination, which is both misleading and extremely frustrating.

Thus, the most important claim is to show that one's results are reproducible and that phylogenetic analysis of the sequences is consistent with the authenticity of the sample. This involves both extraction of several samples from one individual specimen and reproduction of the experiment in different laboratories.

8.2 THE SCOOP ON CONSERVATION

We suggest that many of the techniques for studying ancient DNA can be successfully applied to modern endangered species or populations. Pääbo's group has used these techniques to study conservation genetics of the European brown bear, *Ursus arctos*, which exists in large populations in Scandinavia and Russia, and in a few smaller populations in western Europe. The smallest European brown bear population exists on the southern slopes of the Alps in Brenta, Italy. Here, very few brown bears compose the last indigenous population of bears in the Alps. Sightings of bears have drastically declined, and there may now be fewer than ten animals in this population.

A goal of the provincial government was to study the genetics of this population in preparation for a restocking effort. Traditionally, techniques for collecting DNA — usually by tranquilizing individuals and taking blood samples — while not physically harmful are still invasive and stressful to the animals. Infrared cameras allow one to observe them at night, but the government needed an approach that would not disturb these very few remaining animals. One could sample hairs from the trees where the bears scrub themselves, but such locations are hard to find, and it is also difficult to determine whether the samples have been collected from several individuals or just one. Michael Kohn in Pääbo's laboratory suggested that they could collect DNA from other traces that bears leave behind; for example, their scatological remains. He sampled droppings from individual bears and extracted DNA by using a silica resin that removes many inhibitory components. The resulting "excremental PCR" experiment produced a 150 bp mitochondrial DNA fragment in more than half of the extracts. Taking several extracts from one particular sample also increased the chance of isolating DNA that was suitable for PCR. Besides sequencing approximately 350 bp from the variable mitochondrial control region, in some cases Kohn *et al.* (1995) also determined the sex of each bear by amplifying a fragment from the single-copy testis-determining-factor gene (*SRY*) on the Y chromosome. This produces an amplified fragment only in males. Including the mitochondrial PCR primers in the same PCR as a control produces a single mitochondrial band for females and a doublet (both mitochondrial and Y-chromosomal bands) for males.

A comparison of the sequences from the Brenta bear population with homologous sequences from all other populations in Europe revealed that the Brenta population actually contains just one mitochondrial allele. This fact alone was not surprising, given the small size of the population, but this particular allele is shared with a single population in Slovenia. Phylogenetic analysis also revealed a deep split in the mitochondrial gene pool that separates brown bear populations in Russia, Slovokia, northern Sweden, and Romania from populations in Western Europe, southern Sweden, Spain, and other western populations. Thus although these populations do not appear morphologically distinct, they are apparently genetically isolated. Even in Romania, where bears containing sequences from both clades exist side by side, sequences from the "eastern" and "western" clades within a single locality differ as much as the entire European population. So in the plan to restock the Brenta population, it would be unwise to introduce bears from Finland, because these studies showed that the eastern and western types are genetically different. Instead, the reintroduction of bears from Slovenia into Brenta would allow a mixture of the Brenta allele with others found in Slovenia.

Other informative sequences can also be recovered from DNA in droppings. For example, primers specific for the chloroplast *rbc*L (ribulose-1,5-biphosphate carboxylase) gene can detect plant sequences that allow one to study an individual herbivore's diet (Höss *et al.* 1992). This study revealed seasonal variation in the diet: during the fall, when a particular plant's berries apparently constitute most of the bears' diet, one plant (*Photinia villosa*) is the predominant source of amplified chloroplast DNA. Thus, such a noninvasive approach allows the study of not only the genetics but also several aspects of the ecology of these individuals, beyond the information one can learn by morphological examination of the scats. The future of "molecular scatology" has much to offer conservation biologists.

8.3 MUSEUM SPECIMENS: A CENTURY-LONG POPULATION SURVEY

Over the last few years, the techniques for studying ancient DNA have been most successfully developed for and applied to museum collections: dry skin and bones of specimens collected over the past century. It is often straightforward to retrieve DNA from such remains, and it generally produces a sequence of very good quality. A second advantage of working with recent remains is that one can compare museum specimens to present-day populations, often taken from the same locations 50 to 100 years earlier. This has potentially profound implications for conservation genetics, as one can experimentally test the decline in genetic variation within endangered populations over the course of time. Kelley Thomas, Francis Villablanca, and Svante Pääbo performed such a study in Allan Wilson's laboratory in 1990. They compared forty-three specimens of the Panamint kangaroo rat, *Dipodomys panamintinus*, in the Berkeley Museum of Vertebrate Zoology from three geographically distinct subspecies in California and compared these to sixty-three specimens collected from the same locations in 1988 (Thomas *et al.* 1990).

They found that the three populations had actually been very stable, with maintenance of genetic variation, and that the temporally displaced mitochondrial lineages grouped together consistently over time. Even the subspecies with the smallest population preserved most of its variation in mitochondrial types, despite the highest expected probability of change of mitochondrial gene frequencies, due to the influence of drift or immigration from neighboring populations. Also, the least amount of new variation was introduced into this population between 1917 and 1988. The same population contained a mitochondrial type in 1988 that was shared with a lineage in another small population (Thomas *et al.* 1990). While the conventional interpretation might be that this reflects a recent introgression from one population into another, DNA sequences from museum specimens document the presence of the very same mitochondrial lineage in this population in 1911. Hence, either the immigration was not recent, perhaps dating to a time when these subspecies had overlapping geographic ranges, or alternate explanations, even selection, could be responsible for the maintenance of this mitochondrial allele at a low frequency in two separate populations. Thus, museum collections can provide an important source of information about several temporal aspects of potentially endangered populations. Such questions may be critical in determining the influence of recent human activity versus long-term natural causes on levels of variation in natural populations.

8.4 Gleaning Ivory and Black Caviar

Conservation forensics uses genetic information to enforce laws that protect endangered species. Just as criminal forensics uses genetic screening to aid identification, the recovery of DNA from elephant tusks and attached bits of dried tissue (Georgiadis *et al.* 1994) has been investigated as a source of identification of the geographical origin of elephant products. Such information would help to enforce strict poaching laws in Africa (Cherfas 1989; Templeton 1991).

During the ban on all ivory trade between 1989 and 1997, elephant populations in Zimbabwe, Botswana, and Namibia did increase, but poaching, particularly in other countries, continued, and the overall population still decreased, though less abruptly: populations of the African elephant, *Loxodonta africana*, dropped from 1.2 or 1.3 million in 1979 to 625,000 in 1989 and then to 580,000 during the ban (Lemonick 1997). Since culling of elephants was legal in Zimbabwe, Namibia, and Botswana, these governments amassed nearly 100 tons of ivory, worth

approximately eight billion dollars in 1997. Some of this bounty came
from animals that died naturally or were threatening people, while some
came from the confiscation of illegally obtained ivory over the seven-
year ban. Zimbabwe's practices channeled money for elephant products
to impoverished residents whose crops and sometimes even lives faced
destruction by marauding elephants. By applying the principle of sustainable use to African elephant populations, the other governments as
well hoped that controlled limited trade would encourage expansion of
the elephant's range and numbers, as it had for Zimbabwe. In other
words, all the governments expect that increasing the economic value
of *L. africana* would maximize the local incentive to conserve both the
species and its habitat.

An ongoing problem is the discrimination between legal ivory and
contraband. Hunting of elephants is still illegal in countries north of
Zimbabwe where poaching has had the greatest impact. An elephant's
diet leaves a signature of stable isotopes in its tusks, but this provides an
expensive means of identifying ivory's origins. John Patton and Nicholas
Georgiadis thus investigated the possibility of recovering DNA from
tusks (Cherfas 1989; Templeton 1991). Like DNA from bone, this DNA
is typically degraded; however Templeton (1991) reported the detection
of PCR amplified fragments from nuclear and mitochondrial DNA. More
accessible, the DNA in associated dried tissue has yielded fragments as
long as 2,450 bp, and Georgiadis *et al.* (1994) included DNA from scrap-
ings of several tusks in a survey of mitochondrial DNA restriction site
polymorphism within 270 individuals in savanna elephant populations.
In addition to providing information about the current population structure and historic patterns of gene flow, such studies will provide a set of
signature haplotypes that may be helpful in assigning DNA samples to
particular geographical origins to identify legal or illegal ivory.

In a similar vein, a number of other studies have forced PCR into
the limelight of conservation forensics. For example, Baker and Palumbi
(1994) identified the sale of humpback, fin, and minke whale meat in
Korean and Japanese markets by phylogenetic analysis of amplified mitochondrial DNA fragments. Amato and colleagues (1993, 1998) used a
PCR approach to examine the trade of caiman crocodile skin remains in
commercial clothing and accessories. Later, DeSalle and Birstein (1996)
developed a PCR assay based on mitochondrial sequence variation that
identified misrepresentation of species on labels of canned caviar. Many
species of sturgeons and paddlefishes are threatened by unregulated over-
fishing plus habitat destruction, and the latter study identified the sale
in Manhattan of black caviar from threatened species of sturgeon, one
of which has already become extinct in the Aral Sea.

Putative Age of DNA	Source	Reference
120–135 Myr*	weevil in amber	Cano et al. 1993
80 Myr	possible dinosaur bone fragment	Woodward et al. 1994
35–40 Myr	extinct plant leaf in amber	Poinar et al. 1993
25–40 Myr	stingless bee in amber	Cano et al. 1992
25–40 Myr	bacterial spores in amber	Cano and Borucki 1995
25–30 Myr	extinct termite in amber	DeSalle et al. 1992
17–20 Myr	magnolia leaf compression	Golenberg et al. 1990
17–20 Myr	bald cypress leaf compression	Soltis et al. 1992
30,000–100,000 years	Neandertal bone	Krings et al. 1997
19,600–20,000 years	extinct ground sloth in dung (coprolite)	Poinar et al. 1998
19,600–20,000 years	seven Pleistocene plants in coprolite	Poinar et al. 1998
10,000–53,000 years	extinct woolly mammoth	Johnson et al. 1985
14,000 years	saber-toothed cat in La Brea tar pit	Janczewski et al. 1992
13,000 years	extinct ground sloth (bone/soft tissue)	Höss et al. 1996a
5,000 years	Tyrolean Ice Man	Handt et al. 1994b
3,300 years	New Zealand moas	Cooper et al. 1992
2,400 years	Egyptian mummy	Pääbo 1985
140 years	Extinct quagga	Higuchi et al. 1984
120 years	Extinct marsupial wolf	Thomas et al. 1989

* Myr = millions of years

Table 8.1. Representative studies that reported ancient DNA.

8.5 EXTINCT SPECIMENS: GONE BUT NOT FORGOTTEN

Much can be learned from the study of animals that are already extinct. Table 8.1 lists numerous reports in the recent literature that claimed recovery of ancient DNA from extinct animals, whether preserved in museum specimens, compression fossils, bone, or amber (reviewed in Austin *et al.* 1997). There has been much debate about the authenticity of such claims, and many have failed the test of reproducibility in other labs, despite often reasonable phylogenetic groupings of the resulting sequences (Austin *et al.* 1997; Walden and Robertson 1997) and even the most incredible claims to have revived bacterial spores preserved in 25- to 40-million-year-old Dominican amber (Cano and Burucki 1995). Museum specimens continue to provide the most reliable source of ancient DNA for the study of endangered or extinct species.

One illustrative example concerns the moas of New Zealand, which are an extinct group of large, flightless birds that coexisted with the kiwis in New Zealand during the Pleistocene. The moas probably had a great influence on the flora and fauna of New Zealand, as there were apparently very few animals. For example, eleven different types of plants in New Zealand may have independently evolved a defense against moa grazing by protecting their leaves that now grow inside the plants.

The moas were part of a larger group of flightless birds, the ratites, that typically exist on large landmasses. These include ostriches in Africa and formerly in Asia; rheas in South America; the emu and cassowaries in Australia and Papua New Guinea; and three extant species of kiwis in New Zealand. All are related to the tinamous in South America: birds with limited capacity for flight.

The classical view of the evolution of this group of birds is that they exist on southern landmasses because an ancestral species became isolated on ancient Gondwanaland, and then the major ratite lineages diverged following the breakup of Gondwanaland. Hence, it was presumed that an ancestor of both the kiwis and the moas arrived in New Zealand via the land bridge that used to exist between Antarctica and Australia until about 80 million years ago. This scenerio predicts that the closest living relative of the extinct moas would be the kiwis that exist on New Zealand today.

To test this hypothesis, Alan Cooper (Cooper *et al.* 1992) obtained samples from four of the eleven species of moas that have been identified by paleontologists, including both soft tissues and bones from 3,000-year-old mummified remains in caves. In cases in which both types of tissue were available, longer amplifications of DNA, up to 400 bp, were typically more successful with DNA from bone than from soft tissue such as skin. This indicates that the DNA in bone was better preserved.

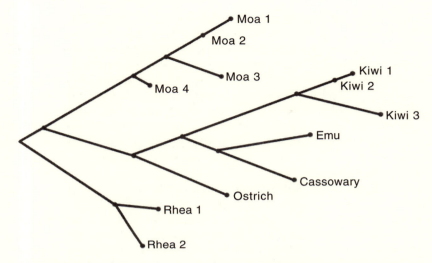

Figure 8.3. Representative unrooted maximum-likelihood tree of ratite birds. Branch lengths are proportional to expected number of substitutions (adapted from Cooper *et al.* 1992).

The assembly of longer sequences from overlapping PCR fragments provided enough sequence information to reconstruct a phylogenetic tree comparing the four moa species to other ratite species (Fig. 8.3).

As expected, Cooper *et al.* observed that the moa species were all closely related to each other, as were the three extant species of kiwis on New Zealand, but these data surprisingly revealed that the moas and the kiwis are not closely related to each other. The moas instead represent an early diverging lineage among this group of birds, whereas the kiwis are actually more closely related to the other ratites, including ostriches, Australian emus, and cassowaries, than they are to the moas. Therefore, while it is plausible that the moas became isolated on New Zealand 80 million years ago, this result suggests that the kiwis arrived much later, because they are more closely related to Australian birds. Nonetheless, the fact that all of these birds are flightless begs the question of their mechanism for dispersal. An unlikely possibility is long-distance swimming. More likely, as the fossil record is consistent with a scenario in which the ratites evolved from an ancestor that could fly (Houde 1986), several lineages may have independently lost their capacity for flight after reaching New Zealand, which offered no ground predators.

In addition, recent studies of other plants and birds have suggested that the most closely related taxa to ones found in New Zealand often diverged 40–50 million years ago from lineages found in Australia

(Marchant and Higgins 1990; Christidis and Boles 1994; Cooper and Penny 1997). Hence, it appears that a few instances of island hopping may have permitted species to travel from Australia to New Zealand until approximately 40 million years ago, possibly over the New Caledonia bridge. Dramatically underscoring this observation was the recent discovery of a dinosaur jawbone in New Zealand. This disproved the long-standing belief that dinosaurs had never been native to New Zealand because it was too isolated. This also takes us the next step backwards in time.

8.6 THE GLITTER OF AMBER

Even before Michael Crichton wrote *Jurassic Park*, conservation and evolutionary biologists had asked the question of whether one could retrieve DNA sequences from fossil remains to study species evolution over millions of years rather than thousands of years, as has been the focus of our earlier discussion. This would wonderfully allow the study of species that have long been extinct. Indeed, the retrieval of DNA from amber-preserved specimens has permitted the reconstruction of molecular phylogenies containing extinct taxa.

In one particularly illuminating example, DeSalle *et al.* (1992) challenged the classical view of insect evolution that termites evolved from cockroaches by sequencing portions of both the nuclear and mitochondrial ribosomal RNA genes from a 25- to 30-million-year-old fossil termite, *Mastotermes electrodominicus*. Represented today by only *Mastotermes darwiniensis*, the Mastotermitidae were considered a separate family and potentially a missing link between cockroaches and termites, because they share some primitive morphological features with cockroaches. However addition of the *M. electrodominicus* sequence to this tree revealed that *Mastotermes* is actually a separate genus within the termites, which are resultingly a monophyletic clade (Fig. 8.4).

Thus, this careful study overturned the original hypothesis that cockroaches graded into termites via a *Mastotermes*-like intermediate (DeSalle *et al.* 1992; DeSalle 1994). In addition to many careful negative controls, the authors of this study took the most strident measures to perform DNA extractions from amber in a brand-new laboratory where no insect DNA had been studied, thus minimizing the risk of using contaminated equipment or reagents. In addition, most sequence contaminants when identified were generally drosophilid in origin. Sequence comparisons of the *M. electrodominicus* DNA confirmed that it was most closely related to *M. darwiniensis*, consistent with the expected relationship between the fossil and living *Mastotermes*, but there were

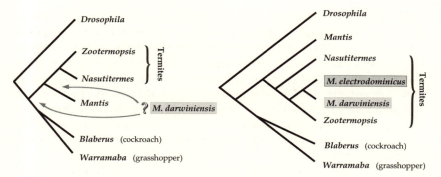

Figure 8.4. Molecular phylogeny of termites and cockroaches. Addition of the fossil sequence for *M. electrodominicus* (on the right) resolves the controversial placement of *M. darwiniensis* on the tree on the left (adapted from DeSalle 1994).

also sufficient sequence differences to imply that the fossil sequence was unique and not a contaminant from *M. darwiniensis*.

Amber-preserved specimens are generally completely entombed in the resin, which offers exquisite preservation of morphological detail. Moreover dehydration preserves tissues in the same manner as mummification (Grimaldi *et al.* 1994), and the terpenoids in amber may act as an antimicrobial agent (Langenheim 1990; Austin *et al.* 1997), slowing down the process of decay. Nonetheless, no claim to have isolated DNA from insects preserved in amber has ever been successfully reproduced in another laboratory, although at least four have tried (Howland and Hewitt 1994; Pawlowski *et al.* 1996; Austin *et al.* 1997; Walden and Robertson 1997). These results are troubling and challenge the authenticity of all claims to have recovered ancient DNA from amber.

8.7 Dinosaur DNA or Mine?

In 1994, Woodward *et al.* published a report in *Science* that they had recovered DNA sequences from 80-million-year-old bone fragments, which morphologically must have come from a dinosaur, although the species was unknown. From these bones in Utah, they reported eight different sequences altogether recovered by PCR with primers designed to amplify the cytochrome *b* gene in the mitochondria. The sequences are quite (up to 10%) different from each other and also from everything else in the database, as of 1994. However, phylogenetic analysis of these sequences revealed that they are not associated with birds as one would predict for a dinosaur sequence, but rather they cluster more often with human sequences. On the other hand, the sequences are certainly not human cytochrome *b* as they are almost 30% different from the human sequence. Pääbo and others offered an explanation that such sequences could be human mitochondrial pseudogene sequences that had been transferred from the mitochondria to the nucleus, where they would have accummulated a series of mutations since their divergence from the parent mitochondrial sequence. Such a suite of events would make these genes look different from everything else but still retain their resemblance to human cytochrome *b*.

To test this hypothesis, Zischler *et al.* (1995) isolated nuclear DNA that is free from mitochondrial DNA by releasing the heads of human sperm. As the mitochondria in sperm are found only on the tails to provide energy for swimming, digestion of the midpiece with (spermicidal) detergents allows enrichment of the heads on sucrose gradients. A Southern blot with mitochondrial DNA probes verified that the sperm DNA preparations contained pure nuclear DNA. However, at lower stringency, even the highly purified nuclear DNA still contained bands that hybridized to the mitochondrial probe. These bands, Zischler *et al.* concluded, are presumably derived from mitochondrial-like sequences in the nucleus. Subsequent PCR amplification of the purified nuclear DNA using the same primers as those described in Woodward *et al.* (1994) produced two sequences that grouped consistently in phylogenetic trees with the sequences originally reported from Utah bone fragments.

All sequences that Zischler *et al.* obtained clustered with human cytochrome *b* as well as with the putative dinosaur sequences. Thus, given the incredulity of alternative hypotheses, such as contamination of human nuclear DNA by dinosaur DNA, admixture of human and dinosaur populations, or peculiar lateral transfer of genes in the late Cretaceous, the most likely explanation for these data is, of course, human contamination of the dinosaur bone specimens. This example highlights one of the challenges of working with ancient DNA, as human pseudogenes provide a persistent and pernicious source of DNA contamination (Collura and Stewart 1995). Even the most carefully designed controls will rarely detect such unexpected sources of contamination.

8.8 DNA Damage versus Preservation

Finally, we ask if there is a good way to identify ancient remains that are most likely to contain DNA. Some processes of DNA degradation are actually independent of time. Experimentally the preservation of DNA is very similar whether from a 13,000-year-old ground sloth from

Chile (Höss *et al.* 1996a), 2,000- to 4,000-year-old human remains from ancient Egypt and South America, a 120-year-old marsupial wolf from a zoological collection (Thomas *et al.* 1989), or a 4-year-old piece of dried cured salami. Hence, DNA fragmentation to short pieces must occur quickly after death, probably by damage at nucleosomal sites.

Other types of DNA damage do accumulate over time. One can survey the damaged bases in DNA by acid hydrolysis to release the free bases, which can be separated by chromatography. Such an analysis of modern DNA will yield the four standard bases plus some methylated cytosine. Ancient samples (1,000 years old or older) produce approximately the same relative amounts of purines (G's and A's), but the pyrimidines (C's and T's) are considerably underrepresented, often less than 5% of the expected amounts. Pyrimidines are very sensitive to oxidation, and hence modified pyrimidines are abundant in archaeological remains.

An analysis using gas chromatography/mass spectrometry revealed the identity of several modified bases in ancient remains (Höss *et al.* 1996b). Hydantoins, which are modified pyrimidines, are the most common compounds. These form as a result of oxidative base damage, caused by either direct irradiation or free-radical attack of DNA in aqueous solution. Significantly, the level of hydantoins seems to have an inverse correlation with the ability to amplify DNA from ancient remains. As these bases have no appropriate base-pairing partners, such lesions are an effective barrier to strand elongation during DNA polymerization, which inhibits PCR. Other types of oxidative damage to purines, producing 8-hydroxyguanine for example, generate miscoding lesions in the DNA. In such an instance, the polymerase incorporates an incorrect base but is able to extend the template. Thus, one would not expect these lesions to reduce the efficiency of PCR, and correspondingly they do not correlate with the inability to amplify ancient DNA. However, the presence of such miscoding lesions does reduce the accuracy of individual amplified clones.

Another conclusion of this study (Höss *et al.* 1996b) was that sample preservation appears to be best at low temperatures, such as in arctic and subantarctic regions. Such conditions, or even freezing, presumably decrease the rate of postmortem DNA degradation. This suggests that sample retrieval from cold climates is optimal whenever one wishes to recover DNA from biological remains. However, since gas chromatography/mass spectrometry analysis requires at least one μg of DNA — or a few grams of bone or tissue — it is hardly feasible in cases where sample material is limiting.

8.9 SAVED BY ASYMMETRY

At the beginning of this chapter we remarked that only 2 of 110 Egyptian mummy samples surveyed actually gave reliable results. With such a small success rate, the experimenter must screen large numbers of specimens just to retrieve a small number of sequences. This is especially unsatisfactory when the screening of archaelogical remains often entails destruction of the fossil itself. In addition, a better understanding of the fossilization conditions that are likely to preserve DNA would aid in the choice of specimens that are likely to yield positive results and obviate the need to perform thousands of DNA extractions and PCR reactions. The brute-force-screening approach also does not recognize a contaminant from an authentic sample before sequencing, and hence several sequencing reactions are resultingly wasted on contaminating DNA. Jeffrey Bada originally suggested that amino acid preservation might correlate with the state of DNA preservation and thus provide a criterion for determining the likelihood of preservation of endogenous DNA.

Nineteen of our twenty amino acids (all but glycine) have a chiral center and exist in two stereoisomers, or enantiomers, of each other: laevorotatory, L, and dextrorotatory, D, amino acids. Protein synthesis on earth uses only L-amino acids, and even carbonaceous chondrite meteorites that contain amino acids show a detectable excess of L-amino acids (Engel and Macko 1997). Slowly after death, a spontaneous racemization reaction converts the L-enantiomer to the D-enantiomer until it eventually reaches 50%, usually after several thousand years. The rate of racemization differs for each amino acid, but it depends on several variables that also influence the state of preservation of DNA, such as temperature, pH, and metal ion concentrations. Aspartic acid, with one of the highest rates of racemization, conveniently undergoes this reaction at a rate very similar to that of depurination of DNA (the loss of A and G bases from DNA); this is the most common path of DNA degradation in aqueous solution. Therefore, Poinar *et al.* (1996) investigated the ratios of D- to L-aspartic acid in two sets of different archaeological remains: one set contained samples from which authentic DNA sequences had been retrieved, and the other set contained samples from which no DNA sequences could be amplified, despite several attempts.

The first set of specimens had to meet several criteria of authenticity, including being substantiated by experiments in other laboratories. Poinar *et al.* thus excluded human samples from this part of this study because of the difficulty of ruling out modern contamination. The only exception to this was the study of the 5,000-year-old Tyrolean Ice Man, which has been reproduced in other laboratories (Handt *et al.* 1994b).

This set of "reliable" samples all contained predominantly L-aspartic acid and displayed very little racemization to the D-form. Other remains that did not produce successful amplifications, such as a 20,000-year-old horse preserved in a warm climate (southern Chile), contained as much as 30%–40% D-aspartic acid, or a D/L ratio between 0.4 and 0.7.

In general, only the samples that had a very low amount of racemization, with a D/L-aspartic acid ratio below a threshold of 0.1 or 0.08, contained amplifiable material. On the other hand, the Neandertal skull from which Krings et al. (1997) obtained mitochondrial DNA contained a D/L-aspartic acid ratio between 0.11 and 0.12, within the range compatible with some DNA damage but also some DNA survival. In the study of Poinar et al. (1996), all seventeen samples that failed to produce any amplified DNA had D/L ratios greater than 0.1. By extrapolation from the racemization rates of aspartic acid preserved in bone from different climatic regions, Poinar et al. conclude that DNA does survive marginally longer in cold regions than in warm regions (10,000 years versus a few thousand years), and later studies indicated that permafrost and Holocene samples are indeed the most likely to contain DNA (Cooper et al. 1997). Furthermore, Poinar et al. observed some correlation between the extent of racemization and the maximum length of the DNA fragments that could be amplified (Fig. 8.5; Poinar et al. 1996).

Therefore the D/L-aspartic acid racemization assay, which can be performed on small (1mm or 2mg) sections of tissue or bone, may indeed provide a facile method to screen for samples that are likely to contain ancient DNA. The 20-minute procedure could even be automated as it only requires sample hydrolysis (which itself produces racemization of approximately 5% of the material) followed by High-Performance Liquid Chromotography (HPLC) analysis. In addition, the screening of associated less important remains, such as animal teeth and bones found with the fossil of interest, can obviate the need to destroy part of an important fossil for analysis. Such associated remains provide a suitable proxy for determining the likelihood of DNA preservation in material from the same location (Cooper et al. 1997). Furthermore, since aspartic acid undergoes racemization more rapidly than other amino acids, higher levels of racemization of amino acids such as alanine or leucine also imply the presence of amino acids of different ages resulting from contamination (Poinar et al. 1996).

Poinar et al. (1996) then examined a series of specimens that were several millions of years old. They first performed a careful analysis of four dinosaur bones, including the Utah bone from which DNA sequences had been reported (Woodward et al. 1994), a Tyrannosaurus rex bone from Montana, which preserved some cellular structure (Schweitzer et al.

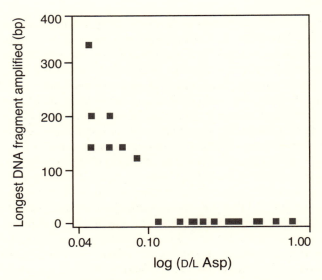

Figure 8.5. The length of the longest amplified DNA fragment depends on the extent of Asp racemization (reproduced with permission from Poinar *et al.* 1996. Copyright 1996 American Association for the Advancement of Science).

1997), and two dinosaur bones that had been preserved under extended cold conditions in Antarctica. All, however, had a D/L-aspartic acid ratio greater than 0.15 and more racemization of alanine than aspartic acid, together suggesting that it is unlikely that such remains would contain any dinosaur DNA.

The next group of specimens they examined were a set of plant fossils formed as leaf compressions on the bottom of a lake from a 17-million-year-old Clarkia deposit in northern Idaho. These fossils contain magnolia leaves among other leaves that are still partially green when the clay layers are split open, and that then turn black rapidly upon oxidization. There has been an extended debate about the authenticity of 800 and 1,200 base pair chloroplast DNA sequences that two groups have reported from such remains (Golenberg *et al.* 1990; Soltis *et al.* 1992).

The analysis by Poinar *et al.* of four leaves from Clarkia revealed disappointingly low levels of amino acids in these remains, with barely detectable levels of aspartic acid in particular. However both D- and L-alanine were present, with a D/L ratio greater than 0.15. Recalling that alanine racemizes even more slowly than aspartic acid, this indicates extensive racemization. Hence, any DNA is likely to be very degraded. Moreover, the finding of some DNA of high molecular mass in fossil plant extracts does not appear to correlate with the ability to amplify ancient plant DNA and most likely derives from bacterial DNA (Sidow *et al.* 1991); hence an artificially high yield of so-called "fossil" DNA may represent a mixture of endogenous and, sometimes overwhelmingly, contaminating DNA.

For an 800 base pair DNA molecule in aqueous solution at 15°C and neutral pH, complete depurination will result in the rapid breakdown of 10^{12} DNA molecules after 5,000 years. Even an 80 base pair molecule, possibly the shortest fragment one would reasonably expect to amplify, decays in approximately 50,000 years, suggesting an absolute limit for the recovery of DNA in the presence of any water. Amber inclusions do offer anhydrous preservation conditions, and claims of ancient DNA from samples preserved in amber over 100 million years have been reported (Cano *et al.* 1993). The morphological preservation of insects, plants, and sometimes even vertebrates in 25- to 40-million-year-old Dominican amber is certainly inspiring. Amino acid analysis of amber inclusions of different ages up to 35 million years old revealed surprisingly low aspartic acid racemization: below 0.1 and in some cases not higher than the fraction induced by the analysis. Therefore, this suggests that amber inclusions might indeed be an optimistic source for retrieving ancient DNA sequences, and this certainly suggests that tests for amino acid racemization should be the first step in any studies attempting to retrieve ancient DNA from amber inclusions. The development of better technology for the recovery and repair of damaged DNA will be critical to the success of this field, especially since reports of obtaining such sequences still have not met the criterion of being reproduced by different laboratories studying the same specimens.

8.10 Conclusions: Looking to the Future Using Glimpses Deep into the Past

Thus, although the chances of recovering dinosaur sequences appear very weak with known forms of preservation, the future of ancient DNA grows stronger. Its applications to conservation genetics are varied, as the tools that allow recovery of DNA from museum specimens permit both the genetic analysis of extinct species — such as the New Zealand fauna — and long-term population studies spanning a century or more. Detailed investigations into the molecular genetics of extinct species may reveal features important to the extinction process itself, such as the loss of variation or beneficial alleles or, conversely, the spread of pathogens or deleterious alleles that may have contributed to a population's or species' demise. For example, the identification of *Mycobacterium tuberculosis*

DNA in mummified tissue from pre-Columbian Peru resolved a contro-
versy surrounding the age of introduction of human tuberculosis into the
New World (Salo *et al.* 1994). This study hints at the myriad possible
inquiries in the field of molecular epidemiological archaeology.

In our well-equipped arsenal of techniques to study the molecular genetics of species past and present, the racemization test — one of the
field's most salient contributions — allows the screening and discovery
of novel sources of DNA. This will, in turn, yield new insight and tools
for the recovery of DNA from unexpected sources. Expanding our knowledge of the deep biosphere, amino acid and PCR analysis of rocks from
the ultradeep gold mines in South Africa has already revealed the presence of nonracemic amino acids and probably extant microbial communities in these extreme environments, 3,200 meters below the surface
(Onstott *et al.* 1997). Such tools are also of key importance in the
scrutinization of possible fossil evidence for ancient life in material of
exobiological origin, such as Martian meteorite ALH84001 (Bada *et al.*
1998; McKay *et al.* 1996). The recovery of DNA and particularly amino
acids from unusual sources thus provides a tool for the discovery of possibly the oldest forms of life on Earth (or Mars), in addition to providing
a safe and noninvasive means to probe several aspects of today's endangered species and populations.

Lastly, the challenge of reconstructing ancestral sequences has been
met not only by the recovery of intact DNA molecules from ancient remains — such as museum specimens or insects in amber — but also by
phylogenetic reconstruction from modern sequences. Careful phylogenetic analysis of present-day sequences can allow accurate extrapolation
to the ancestral molecules. Using modern cloning tools of molecular biology, two groups have actually synthesized active ancestral molecules
in the laboratory (Jermann *et al.* 1995; Messier and Stewart 1997).
These experiments allow exploration of the biochemistry as well as the
genetics of even long-extinct species. Indeed, the synthesis of functional
ancestral proteins inferred from present-day phylogenies provides strong
support for the accuracy of several methods for phylogenetic reconstruction. Though extinct for over three billion years, the biochemistry of the
earliest living cells that gave rise to all life on earth is now a tangible
problem (Benner *et al.* 1993). Ideally, the genome sequencing projects
for several representatives of all three domains of life will provide enough
information to reconstruct the evolution of metabolism over all natural
history, as recorded in our genes. Possibly, one day the reconstruction
of partial genome sequences of long extinct organisms will even be a
reality.

Acknowledgments

Thanks to Svante Pääbo whose talk at Princeton University ("DNA from Unusual Sources in Conservation and Evolutionary Biology") at this symposium formed part of the inspiration for this work; the views contained here are those of the author, who was supported in part by a Burroughs Wellcome Fund New Investigator Award in Molecular Parasitology. The figures are reproduced with permission. The author also wishes to thank Rob DeSalle for helpful comments on the manuscript.

References

Amato, G. D., Gatesy, J., and Brazaitis, P. (1998). PCR assays of variable nucleotide sites for identification of conservation units: an example from Caiman. In DeSalle, R. and Schierwater, B. (eds.), *Molecular Approaches to Ecology and Evolution*, 177-189, Basel: Birkhauser Verlag.

Amato, G. D., and Wharton, D. (1993). A systematic approach to identifying units of conservation: Examples of progress and problems. *Proc. Ann. Conf. Am. Assoc. Zoo. Parks Aquariums*, WR 241, 83–87.

Austin, J. J., Smith, A. B., and Thomas, R. H. (1997). Palaeontology in a molecular world: The search for authentic ancient DNA. *Trends. Ecol. Evol.*, *12*, 303–306.

Bada, J. L., Glavin, D. P., McDonald, G. D., and Becker, L. (1998). A Search for Endogenous Amino Acids in Martian Meteorite ALH84001. *Science*, *279*, 362–365.

Baker, C. S., and Palumbi, S. R. (1994). Which Whales Are Hunted? A Molecular Genetic Approach to Monitoring Whaling. *Science*, *265*, 1538–1539.

Benner, S. A., Cohen, M. A., Gonnet, G. H., Berkowitz, D. B., and Johnson, K. P. (1993). Reading the Palimpsest: Contemporary Biochemical Data and the RNA World. In Gesteland, R. F. and Atkins, J. F. (eds.), *The RNA World*, 27-70, Plainview, New York: Cold Spring Harbor Laboratory Press.

Cano, R. J., and Burucki, M. K. (1995). Revival and identification of bacterial spores in 25- to 40-million-year-old dominican amber. *Science*, *268*, 1060–1064.

Cano, R. J., Poinar, H. N., Pieniazek, N. J., Acra, A., and Poinar, G. O. (1993). Amplification and sequencing of DNA from a 120–135-million-year-old weevil. *Nature*, *363*, 536–538.

Cano, R. J., Poinar, H., and Poinar, G. O., Jr. (1992). Isolation and partial characterisation of DNA from the bee *Proplebeia dominicana* (Apidae: Hymenoptera) in 25–40 million year old amber. *Med. Sci. Res.*, *20*, 249–251.

Cherfas, J. (1989). Science Gives Ivory a Sense of Identity. *Science*, *246*, 1120–1121.

Christidis, L., and Boles, W. E. (1994). The Taxonomy and Species of Birds of Australia and its Territories, Royal Australian Ornithological Union, Melbourne, Australia.

Collura, R. V., and Stewart, C.-B. (1995). Insertions and duplications of mtDNA in the nuclear genomes of Old World monkeys and hominoids. *Nature*, *378*, 485–489.

Cooper, A., Mourer-Chauvire, C., Chambers, G. K., von Haeseler, A., Wilson, A. C., and Pääbo, S. (1992). Independent origins of New Zealand moas and kiwis. *Proc. Natl. Acad. Sci. USA*, *89*, 8741–8744.

Cooper, A., and Penny, D. (1997). Mass survival of birds across the Cretaceous-Tertiary boundary: Molecular evidence. *Science*, *275*, 1109–13.

Cooper, A., Poinar, H. N., Pääbo, S., Radovcic, J., Debénath, A., Caparros, M., Barroso-Ruiz, C., Bertranpetit, J., Nielsen-Marsh, C., Hedges, R. E. M., and Sykes, B. (1997). Neandertal Genetics. *Science* 277, 1021–1024.

DeSalle, R. (1994). Implications of ancient DNA for phylogenetic studies. *Experientia*, *50*, 543–550.

DeSalle, R., and Birstein, V. (1996). PCR identification of black caviar. *Nature*, *381*, 197–198.

DeSalle, R., Gatesy, J., Wheeler, W., and Grimaldi, D. (1992). DNA sequences from a fossil termite in Oligo-Miocene amber and their phylogenetic implications. *Science*, *257*, 1933–1936.

Engel, M. H., and Macko, S. A. (1997). Isotopic evidence for extraterrestrial non-racemic amino acids in the Murchison meteorite. *Nature* 389, 265–268.

Georgiadis, N., Bischof, L., Templeton. A., Patton, J., Karesh, W., and Western, D. (1994). Structure and history of African elephant populations: I. Eastern and Southern Africa. *J. Heredity*, *85*, 100–104.

Golenberg, E. M., Giannasi, D. E., Clegg, M. T., Smiley, C. J., Durbin, M., Henderson, D., and Zurawski, G. (1990). Chloroplast DNA sequence from a Miocene *Magnolia* species. *Nature*, *344*, 656–658.

Grimaldi, D., E. Bonwich, M. Delannoy, and Doberstein, S. (1994). Electron microscopic studies of mummified tissues in amber fossils. *American Museum Novitates*, *3097*, 1–31.

Hagelberg, E., Thomas, M. G., Cook, C. E. Jr, Sher, A. V., Baryshnikov, G. F., and Lister, A. M. (1994). DNA from ancient mammoth bones. *Nature*, *370*, 333–334.

Handt, O., Höss, M., Krings, M., and Pääbo, S. (1994a). Ancient DNA: methodological challenges. *Experientia*, *50*, 524–529.

Handt, O., Richards, M., Trommsdorff, M., Kilger, C., Simanainen, J., Georgiev, O., Bauer, K., Stone, A., Hedges, R., Schaffner, W., Utermann, G., Sykes, B., and Pääbo, S. (1994b). Molecular genetic analyses of the Tyrolean Ice Man. *Science*, *264*, 1775–8.

Higuchi, R., Bowman, B., Freiberger, M., Ryder, O. A., and Wilson, A. C. (1984). DNA sequences from the quagga, an extinct member of the horse family. *Nature*, *312*, 282–284.

Höss M; Dilling A; Currant A; Pääbo S. (1996a). Molecular phylogeny of the extinct ground sloth *Mylodon darwinii*. *Proc. Natl. Acad. Sci. USA*, *93*, 181–5.

Höss, M., Jaruga, P., Zastawny, T. H., Dizdaroglu, M., and Pääbo, S. (1996b). DNA damage and DNA sequence retrieval from ancient tissues. *Nucleic Acids Res.*, *24*, 1304–7.

Höss, M., Kohn, M., Pääbo, S., Knauer, F., and Schröder, W. (1992). Excrement analysis by PCR. *Nature*, *359*, 199.

Höss, M. Pääbo, S., and Vereshchagin, N. K. (1994). Mammoth DNA sequences. *Nature*, *370*, 333.

Houde, P. (1986). Ostrich ancestors found in the Northern Hemisphere suggest new hypothesis of ratite origins. *Nature*, *324*, 563–565.

Howland, D. E., and Hewitt, G. M. (1994). DNA analysis of extant and fossil beetles. In *Biomolecular Palaeontology: Lyell Meeting Volume*. G. Eglinton and R. L. F. Kay, Eds. NERC Earth Sciences Directorate, Special Publication No. 94/1, 49–51.

Janczewski, D. N., Yuhki, N., Gilbert, D. A., Jefferson, G. T., and O'Brien, S. J. (1992). Molecular phylogenetic inference from saber-toothed cat fossils of Rancho La Brea. *Proc. Natl. Acad. Sci. USA*, *89*, 9769–9773.

Jermann, T. M., Opitz, J. G. Stackhouse, J., and Benner, S. A. (1995). Reconstructing the evolutionary history of the artiodactyl ribonuclease superfamily. *Nature*,
374, 57–59 .

Johnson, P. H., Olson, C. B., Goodman, M. (1985). Isolation and characterization of
deoxyribonucleic acid from tissue of the woolly mammoth, *Mammuthus primige-
nius*. *Comp. Biochem. Physiol. B*, *81*, 1045–51.

Kohn, M., Knauer, F., Stoffella, A., Schroder, W., and Pääbo, S. (1995). Conservation genetics of the European brown bear — a study using excremental PCR of
nuclear and mitochondrial sequences. *Mol. Ecol.*, *4*, 95–103.

Krings, M., Stone, A., Schmitz, R. W., Krainitzki, H., Stoneking, M., and Pääbo,
S. (1997). Neandertal DNA sequences and the origin of modern humans. *Cell*, *90*,
19–30.

Langenheim, J. H. (1990). Plant resins. *Am. Sci.*, *78*, 16–24.

Lawlor, D. A., Dickel, C. D., Hauswirth, W. W., and Parham, P. (1991). Ancient
HLA genes from 7,500-year-old archaeological remains. *Nature*, *349*, 785–8.

Lemonick, M. D. (1997). The Ivory Wars. *Time*, 14924, 64–65.

Marchant, S., and Higgins, P. J. (1990). (coordinators) *Handbook of Australian, New
Zealand, and Antarctic Birds*. Vol 1B: Australian Pelicans to Ducks, Melbourne,
Australia: Oxford University Press.

McKay, D. S., Gibson Jr., E. K., Thomas-Keprta, K. L., Vali, H., Romanek, C.
S., Clemett, S. J., Chillier, X. D. F., Maechling, C. R., and Zare, R. N. (1996).
Search for Past Life on Mars: Possible Relic Biogenic Activity in Martian Meteorite
ALH84001. *Science*, *273*, 924–930.

Messier, W., and Stewart, C. B. (1997). Episodic adaptive evolution of primate
lysozymes. *Nature*, *385*, 151–154.

Onstott, T. C., Tobin, K., Dong, H., DeFlaun, M. F., Fredrickson, J. K., Bailey, T.,
Brockman, F., Kieft, T., Peacock, A., White, D. C., Balkwill, D., Phelps, T. J.,
and Boone, D. R. (1997). The deep gold mines of South Africa: Windows into the
subsurface biosphere. *Proc. Internatl. Soc. Optic. Engin. (SPIE)*, *3111*, 344–357.

Pääbo S. (1985). Molecular cloning of Ancient Egyptian mummy DNA. *Nature*, *314*,
644–5.

Pawlowski, J., Kmieciak, D., Szadziewski, R., and Burkiewicz, A. (1996). Attempted
isolation of DNA from insects embedded in Baltic amber, *Inclusion*, Paleontological
Newsletter and Workshop, Cracow, *22*, 13–14.

Poinar, G. O. (1994). The range of life in amber. *Experientia*, *50*, 536–542.

Poinar, H. N., Cano, R. J., and Poinar, G. O., Jr. (1993). DNA from an extinct plant.
Nature, *363*, 677.

Poinar, H. N., Hofreiter, M., Spaulding, W. G., Martin, P. S., Stankiewicz, B. A.,
Bland, H., Evershed, R. P., Possnert, G., and Pääbo, S. (1998). Molecular Co-
proscopy: Dung and Diet of the Extinct Ground Sloth *Nothrotheriops shastensis*.
Science, *281*, 402–406.

Poinar, H. N., Höss, M., Bada, J. L., and Pääbo, S. (1996). Amino Acid Racemization
and the Preservation of Ancient DNA. *Science*, *272*, 864–866.

Roewer, L., Riess, O., and Prokop, O. (1991). Hybridization and polymerase chain
reaction amplification of simple repeated DNA sequences for the analysis of forensic
stains. *Electrophoresis*, *12*, 181–6.

Salo, W. L., Aufderheide, A. C., Buikstra, J., and Holcomb, T. A. (1994). Identification of *Mycobacterium tuberculosis* DNA in a pre-Columbian Peruvian mummy.
Proc. Natl. Acad. Sci. USA, *91*, 2091–4.

Schweitzer, M. H., Marshall, M., Carron, K., Bohle, D. S., Busses, S. C., Arnold, E. V., Barnard, D., Horner, J. R., and Starkey, J. R. (1997). Heme compounds in dinosaur trabecular bone. *Proc. Natl. Acad. Sci. USA*, *94*, 6291–6296.

Sidow, A., Wilson A. C., and Pääbo S. (1991). Bacterial DNA in Clarkia fossils. *Philos. Trans. R. Soc. Lond. B Biol Sci*, *333*, 429–32.

Soltis, P. S., Soltis, D. E., and Smiley, D. F. (1992). An *rbc*L sequence from a Miocene *Taxodium* bald cypress. *Proc. Natl. Acad. Sci. USA*, *89*, 449–451.

Templeton, A. R. (1991). Genetics and Conservation Biology. In Seitz, A., and Loeschcke, V. (eds.), *Species Conservation: A Population-Biological Approach*, 15–29, Basel: Birkhäuser-Verlag.

Templeton, A. R., and Georgiadis, N. J. (1996). A landscape approach to conservation genetics: Conserving evolutionary processes in the African Bovidae. In Avise, J. C. and Hamrick, J. L. (eds.), *Conservation Genetics: Case Histories from Nature*, 398–430, New York: Chapman and Hall.

Thomas, R. H., Schaffner, W., Wilson, A. C., and Pääbo, S. (1989). DNA phylogeny of the extinct marsupial wolf. *Nature*, *340*, 465–7.

Thomas, W. K., Pääbo, S., Villablanca, F. X., and Wilson, A. C. (1990). Spatial and temporal continuity of kangaroo rat populations shown by sequencing mitochondrial DNA from museum specimens. *J. Mol. Evol.*, *31*, 101–12.

Walden, K. K. O. and Robertson, H. M. (1997). Ancient DNA from Amber Fossil Bees? *Mol. Biol. Evol.*, *14*, 1075–1077.

Woodward, S. R., Weyand, N. J., and Bunnell, M. (1994). DNA sequence from cretaceous period bone fragments. *Science*, *266*, 1229–1232.

Zischler, H., Höss, M., von Haeseler, A., van der Kuyl, A. C., Goudsmit, J., and Pääbo, S. (1995). Detecting Dinosaur DNA. *Science*, *268*, 1192–1193.

Index

adaptive radiation, 133, 141–142, 156
adaptive variation, 62
akepa, 143–156
akialoa, 143–155
akiapolaau, 143–156
Allee effect, 6, 8
allelic diversity, 60
amakihi, 125, 130–132, 142–156
amber, 171, 174
   Dominican, 172, 181
American bison, 14
ancient DNA, 82, 94, 163–181
anianiau, 142
Antarctic fur seal, 79
anthropogenic factors, 2
apapane, 124, 143–156
assimilation, 43
avian poxvirus, 133, 151, 153–154, 158

balancing selection, 38
barred owl, 10

Californian Channel Island fox, 81
captive breeding, 48, 130, 158
carrying capacity, 5, 8, 15
Catalina Island, 29
catastrophe, 5, 27
caviar, 170
cheetah, 78
chloroplast DNA (cpDNA), 30, 33, 168, 180
cichlid fish, 10
clutch size, 155
coalescence
   theory of, 37, 109
   time of, 32
   tree of, 116
cockroaches, 174
coconut crab, 114
contamination, 165–166, 176–179
Coreopsis, 31–35, 39–41

creeper, 143–157
Crespis, 31, 35–36, 41
crested honeycreeper, 143–157
critical patch size, 9
crocodile skin, 170
crossbills, 145

damaged DNA, 181
declining-population paradigm, 42
demographic potential, 7
demographic stochasticity, 8
dinosaur, 171, 174–176, 179
   DNA of, 165
dinucleotide repeats, 92, 93
directional selection, 52
dispersal rate, 6, 9, 13, 35
DNA fingerprinting, 75, 81, 83
Drosophila, 47, 63, 64
dusky seaside sparrow, 38, 39

ecomorph, 142–153
economic factors, 2
edge effects, 6, 9
Egyptian mummy, 163, 171, 178
elephant, 169
enantiomers, 178
environmental policy, xiv
environmental stochasticity, 4, 8
European badger, 80
European brown bear, 167
excremental PCR, 167
exotic species, 3, 9
extinction threshold, 7
extinction vortices, 16
extreme environments, 182

Fisher's fundamental theorem, 24
fitness loss, 11, 15
Florida panther, 12
fluctating selection, 52
founder genome equivalents, 54, 56